D0763967

Quality Assurance for Building Design

Quality Assurance for Building Design

Malcolm Taylor

Harry H Hosker

Copublished in the United States with
John Wiley & Sons, Inc., New York

Longman Scientific & Technical,
Longman Group UK Limited,
Longman House, Burnt Mill, Harlow,
Essex CM20 2JE, England
and Associated Companies throughout the world.

Copublished in the United States with
John Wiley & Sons, Inc., 605 Third Avenue, New York, NY 10158

British Library Cataloguing in Publication Data
A catalogue record for this book is available from the British Library

Library of Congress Cataloging-in-Publication Data
A catalogue record for this book is available from the Library of Congress

Set in 10/12 pt Times
Printed and Bound in Great Britain
at the Bath Press, Avon

Contents

Foreword

> Quality Assurance: a means of achieving improved and consistent quality through better management.
>
> There was a need to control quality over the whole production chain; testing the end-product only was too late since it became largely the detection of failure.
>
> Taylor & Hosker, 1992

In an increasingly competitive environment building designers must give clients confidence in the quality of their service and the buildings they produce. To remain healthy they must also provide these services cost effectively. Quality Assurance is an important management tool encouraging efficient design, and for some time a practical guide has been required on the setting up and maintenance of quality systems for designers.

However, for many years the manufacturing industries have been developing Quality Assurance, which has meant that much of the available information has been oriented towards manufacturing and has encouraged the erroneous view that Quality Assurance was irrelevant to the design industry. This guide explains and demonstrates Quality Assurance (or QA) for all involved in building design.

Too often the subject is seen as the circulation of a series of forms, with boxes to be ticked as an end in themselves. But planned QA reduces the time taken to be certain of quality and assists in managing the risks. It frees time for thinking and allows creative energy to be placed where it is most valuable.

This guide clearly shows the place of QA in the building design industry and briefly outlines its historical context; the British, European and international quality standards; and the role of the relevant certification bodies. Reference is also made to the financial and legal implications of QA. The authors then demonstrate stage by stage the subjects to be addressed and detailed methods for setting up and maintaining a quality system. The explanatory information includes many relevant forms and diagrams which are applicable to most, if not all, building design offices.

Following the guidelines offered will not only encourage the development of individual quality systems, but will also save enormous quantities of time in avoiding the reinvention of existing information. The guide concludes with consideration of the international scene and looks towards the future with Total Quality Management.

Malcolm Taylor and Harry Hosker, after years of experience in a multidisciplinary practice, have created a practical and detailed guide. This experience, combined with considerable research, has provided a book of extraordinary detail in a very accessible form.

This is a valuable reference book and practical guide for all involved in building design and clearly demonstrates how to develop, implement and audit QA, demonstrating the potential for the management of quality at all stages. Quality Assurance need not stifle the designer but faces him. Our industry must take up the challenge and the opportunity to manage quality.

Anthony Furlong
Sheppard Robson
London, 1992

Acknowledgements

The authors gratefully acknowledge the unfailing assistance and encouragement which Building Design Partnership has given in the preparation of this book. In particular, we thank the partners for permission to quote from their quality procedures, and for material provided by their office information centres. We also thank colleagues for helpful comment and in particular Brian Philpott and Walter Berry.

The authors also wish to thank National Accreditation Council for Certification Bodies, British Standards Institution Quality Assurance, Construction Quality Assurance, Lloyd's Register Quality Assurance Ltd and Yarsley Quality Assured Firms Limited for their invaluable comments on the accuracy of the text in the areas of accreditation and certification.

We are grateful to the following for permission to reproduce copyright material:

The Association of Consulting Engineers for extracts from the 1983 edition of Agreement 4A(i) of the *ACE Conditions of Engagement* (1981); BSI for extracts from British Standards. Complete copies of the standard can be obtained from BSI Sales, Linford Wood, Milton Keynes MK14 6LE; RIBA Publications Ltd for extracts from The Royal Institute of British Architects 1982 Architect's Appointment (Procurement and Construction 109); the Royal Institution of Chartered Surveyors for an extract from *QS Fee Scale 36.*

We are also indebted to the following for permission to reproduce copyright figures and tables:

The Association of Consulting Engineers for Fig. 9.1; BSI for part of Fig. 4.2 (BSI quality mark); Building Project Information Committee for Figs 13.1–13.3 (CPI); Construction Quality Assurance for part of

Fig. 4.2 (CQA quality mark); Lloyd's Register Quality Assurance Ltd. for part of Fig. 4.2 (LRQ quality mark); the Secretary of State for the Department of Trade and Industry for part of Fig. 4.2 (NACCB quality mark); SGS Yarsley Quality Assured Firms Ltd. for part of Fig. 4.2 (SGS Yarsley quality mark).

Glossary of Terms

This glossary includes terms used by the authors but not defined in the British Standards. Terms used in British Standards are covered in Chapter 3 and elsewhere in the text as appropriate.

Firm

An organization which provides services to another organization. In BS 5750 terminology, the firm would be the 'supplier'. Typically, a firm is a partnership or company providing services to a client. (See below for 'client'.) Alternatively, the firm could be any group or body within an organization which has been identified as providing a discrete service to another body within the system. This could apply to, say, a part of a public authority, where the 'firm' is a design group, providing a service within the much greater range of services provided by the authority as a whole.

Client

The organization to whom services are to be provided under terms of engagement (see below for terms of engagement). In BS 5750 terminology the client would be the 'purchaser'. Alternatively, as above, where the firm is a group providing a discrete service within the organization, the group or body receiving the service would be the client.

Conditions of engagement

The contract between firm and client under which the firm has promised to provide services in exchange for remuneration. In BS 5750 terminology, the conditions of engagement (sometimes referred to as terms of engagement) would be the 'contract'. Alternatively as above, where firm and clients are groups within the organization, terms of engagement (if any) may not have the legal significance of a contract, and may not be so clearly defined. Nevertheless, there should be a defined service which is to be provided, which serves a similar purpose in the operation of the quality system.

Senior principal
The person in the firm who has the authority to implement and maintain the quality system. He formally commits the firm to its quality system and makes the necessary resources available. Where a firm has only one principal, it is axiomatic that he has the necessary authority. Where the firm has partners or directors or is a public authority, they normally nominate one of their number to this role, and give him the necessary authority. Usually the senior principal, managing director, or chief officer will take the role of senior principal, but the body may decide to retain this authority, appointing one of their members as the contact point.

Job
The individual commission or service traditionally provided by the firm to the client. The alternative popular term is 'project'.

Job partner
The partner charged by the partnership to be responsible for the fulfilment of the commission. (Equivalent terms for 'partner' and 'partnership' will exist in companies and public bodies.)

Profession partner
The firm's partner responsible for monitoring the quality of service for the respective profession (in a single discipline firm the profession partner will be the job partner; in a multidiscipline firm one of the profession partners may also be the job partner).

Design team leader
The person appointed by the professions, whose terms of engagement include the co-ordination and integration of the work of other designers, to perform these duties.

Job quality plan
A document which sets out how the job team intends to discharge the firm's responsibilities under the terms of engagement. It includes information about the job.

Profession representative
The person or persons appointed to develop the procedures on behalf of a profession. They may also be charged with advising members of staff on the interpretation of the procedures and acting as a focus for feedback.

QA manager
The person appointed by the senior principal to manage the auditing and review process and reporting to the senior principal. He may also have

other functions such as leading the development and maintenance of the procedures.

Gender

By convention, the male gender has been used throughout this book. Wherever it appears it refers to either the male or the female gender.

Italics

Italics within the text have been used to denote words and phrases drawn from other documents, principally the British Standards.

Introduction

The Place of Quality Assurance in Building Design

Quality assurance may not yet have come of age, but its credentials are now firmly established as a fundamental aspect in manufacturing and design organizations. There are now few business activities which have not been acquainted with its principles. Its boundaries may not yet be finally defined, but one thing is certain; the value of a well conceived quality system is now widely recognized. The current expansion of quality assurance in the UK is some testimony to this statement. There is now a wider awareness of the need for professional management by a nation seen from offshore as adopting instinctive, intuitive (or less kindly, amateurish) attitudes to management. The manufacturing industries have been aware for many years of the important relationship of management to quality, and, with recent government encouragement, have seen quality assurance as a means of achieving better or consistent quality through better management. These concepts are increasingly seen as equally relevant by the service industries.

Within the spectrum of 'services' lies design, and of course particularly for this book, building design. The formality of 'management' has not come as naturally to designers of buildings as it has to manufacturers, who have long recognized the art as a discrete element in their vocabulary. The excitement of creation is fundamental to design; not an obvious bedfellow with management. Yet the size and complexity of construction projects, the diversity of skills (including the diversity of the construction industry itself) and the substantial changes in how the professional designer is seen by society, demand a recognition of the need for sound management if the organization is to survive. Thus it has been accepted that the basic building blocks of management are no different in principle for designers than they were for the defence industry, where quality assurance had its origins.

Scope and Purpose of this Book

Manufacturing industry is relatively well served in the information it needs for the application of a formal quality system. As we have said above, its interest in such systems has been long established, and manufacturing industry has the advantage that the principles and practice of modern quality management grew from defence. Although a Standard has now been written which covers the interests of the service industries, the principal quality British Standards are still largely written around the needs and language of the manufacturer.

The designer will find a wealth of guidance on working methods. The Royal Institute of British Architects have over many years produced excellent material in the furtherance of sound practice management, and indeed have recently produced Guidance for an office manual in quality management. The Association of Consulting Engineers and the Royal Institution of Chartered Surveyors have also produced guidance for their respective professions. Such guidance cannot, as it stands, be used as the firm's procedures. These must be developed by the firm to suit its particular area of practice and methods of working. They must also be written in a way which will enable them and their implementation to be audited. What is missing is a guide on how to do this to meet the requirements of the British Standards. Among other things this book aims to fill this gap.

It has been said that for quality assurance to be entirely successful, all the component of the many services and construction activities which contribute to the finished building should have quality systems which comply with a common standard. While this may be so, it must remain at the moment a counsel of perfection which will not be achieved for some time. Those currently seeking to invest in formal quality systems naturally concentrate (and in the authors' opinion, rightly) on the quality system needs of their own organizations. Such needs are recognized in this book.

This book is intended as a practical approach to quality assurance. At its heart are the actions necessary to develop, apply and maintain a quality system which is right for the particular organization. Every organization is different, with different priorities and different beliefs in what should comprise its quality related processes which have to be documented and applied. A book should not attempt to lay down 'standard' quality documentation which can simply be taken off the shelf and used. However, we have ventured to suggest areas where experience has shown that design practice may benefit from the adoption of a given approach. We also use procedural matter where it is considered helpful in the illustration of a principle. Readers should not assume this advice to be definitive of the needs of the particular organization; in any event the authors would not pretend to have such an encyclopaedic knowledge!

The more peripheral parts of the book, historical, legal, international and total quality management aspects, could no doubt have been considerably expanded, and there is much further material of interest to the reader who wishes to pursue these areas. To have included it, however, would have produced a much longer book, and might have risked diverting attention from its main aims.

Cross-referencing between chapters has been kept to a minimum, so readers with a particular interest should have little difficulty in identifying and confining their reading to those parts of the book which interest them.

Structure

The structure of the book is as follows:

Chapter 1 explains quality assurance in the context of building design.
Chapter 2 describes its historical context.
Chapter 3 reviews the context of the relevant British and International Quality Standards.
Chapter 4 describes the purpose and function of the certification bodies, and the advantages of third party certification.
Chapter 5 explores situations where QA is applicable.
Chapter 6 advises on developing the quality system.
Chapter 7 continues this process by describing how to prepare quality system procedures.
Chapter 8 explains the preparation of the quality system manual and application of the system.
Chapter 9 deals with conditions of engagement.
Chapter 10 examines purchasing and subletting.
Chapter 11 looks at job administration.
Chapter 12 looks at briefing.
Chapter 13, the requirements for design and production information.
Chapter 14 explores procurement and production information.
Chapter 15 looks briefly at advisory work.
Chapter 16 covers support systems.
Chapter 17 explains the importance of the job quality plan.
Chapter 18 reviews the auditing process.
Chapter 19 deals with the maintaining of the quality system.
Chapter 20 looks at financial considerations.
Chapter 21, legal implications.
Chapter 22, the international scene.
Chapter 23 concludes with a look into the future – Total Quality Management.

The Readership

It will be clear from the foregoing that the book is addressed to a specific sector of the construction industry, the designers of buildings. These will comprise design professionals either practising as independent consultants or employed in public authorities or as designers in design and build organizations. Professions will be principally architects, services and structural engineers, construction cost experts, landscape architects, planners, interior designers and all the other professions which may contribute to the design process. The academic world is taking an increasing and welcome interest in quality assurance, so it is hoped that much will be found of interest by teachers, students and researchers.

Many clients interest themselves not only in the finished service they receive, but also in the means by which their consultants control the processes of providing the service. In fact, many clients demand evidence of a formalized quality system. They will find the content of the book of interest. Clients, and, where appointed, their project managers, may find the book helpful, in the application of quality assurance to the project management organization itself and in the exploration of any quality management demands to be made by the project manager of the organizations he is managing or co-ordinating.

The Authors

The authors have been employed by a large multidisciplinary building design organization, having several offices. They have been closely involved in its quality system activities over a long period, and were leading members of the team which developed a formal quality system complying with BS 5750: Part 1. The system has been approved by an accredited certification body. This book reflects their combined experience in this field. Although the firm is large, the principles of good design management are universal and may be applied to the smallest project. It is hoped, therefore, that the reader, whatever his background, will identify with the content of the book as it applies to his interest in quality assurance.

Malcolm Taylor is now a consultant in quality management.

1 What is Quality Assurance?

The definition of 'Quality Assurance' (or 'QA' to which it is inevitably abbreviated) will be discussed further in Chapter 3 (British Standards), but the term has been so misunderstood and misused that it was thought that it should have a chapter of its own. The expression is difficult to come to terms with, whether taken at its face value, or even as it is used in the context on which this book is based. Hardly a promising start to a book which opens with an admission that the very subject eludes definition!

The problem starts with the word 'quality'. One dictionary describes it very sensibly as 'that which makes a thing what it is', a description not far short of how it will be defined later as it should apply to the services provided by those who design buildings. However, the hyperbole of modern advertising (and thus, popular usage) has added the inference of something superior to the competition (particularly when the word is used as an adjective). We are on safer ground with 'assurance', which is defined by the same dictionary as 'confidence; a feeling of certainty'.

So perhaps we should turn firstly to the definitions given in the British Standards on quality where we might expect to find clarification on how the British Standards Institution intended the subject to be understood and used.

BS 4778: Part 1 (quality vocabulary) defines 'quality' in Clause 3.1 as

> The totality of features and characteristics of a product or service that bear on its ability to satisfy stated or implied needs

and notes

1. In a contractual environment, needs are specified, whereas in other environments, implied needs should be identified and defined.
2. In many instances, needs can change with time; this implies periodic revision of specifications.
3. Needs are usually translated into features and characteristics with specified criteria. Needs may include aspects of usability, safety, availability, reliability, maintainability, economics and environment.

(It might perhaps be useful at this point to distinguish between the status of the clauses in British Standards, and their accompanying notes. Those who undertake to have quality systems in accordance with the appropriate parts of the BS should observe that the clauses are the requirements of the Standard whereas the notes in the BS are for guidance or explanation only). The reader may now better understand the definition of 'quality', and it might be reasonable to agree that the words used by the BS coincide with the dictionary definition.

A definition of 'assurance' is not (perhaps mercifully) attempted by the BS, but 'Quality Assurance' is defined in Clause 3.6 as

> All those planned and systematic actions necessary to provide adequate confidence that a product or service will satisfy given requirements for quality

. . . and notes

> 1 Unless given requirements fully reflect the needs of the user, quality assurance will not be complete.
> 2 For effectiveness, quality assurance usually requires a continuing evaluation of factors that affect the adequacy of the design or specification for intended applications as well as verifications and audits of production, installation and inspection operations. Providing confidence may involve producing evidence.
> 3 Within an organization, quality assurance serves as a management tool. In contractual situations, quality assurance also serves to provide confidence in the supplier.

The definitions given in BS 4778 are adopted by BS 5750, which, as we shall see later, as a systems Standard is the driving force of QA, so this definition of 'Quality Assurance' is important.

The reader may well ask at this point whether 'Quality Assurance' means 'assurance of quality', and if not, why not? This might well be the most important opening question for the building designer's initial curiosity about QA. He may look for reassurance that it will not be seen as a guaranty of his service (or, as this term might be expressed for those who sell goods, 'product conformance'). Unfortunately, some manufacturers of products have sought to give the impression that certification body approval that their quality system conforms to BS 5750 confers some kind of guaranty. That is erroneous, as will be made clear in Chapter 4 where we will see that both BSI and NACCB have drawn clear distinctions between QA and product conformance. (Examples of Standards which confer product conformance are kitemarks and BBA certificates.)

The BS definition above of quality assurance gives the designer some reassurance on this point. Such reassurance may be drawn from the phrases *that bear on* in Clause 3.1, and *adequate confidence* in Clause

3.6, although it has to be admitted that they do not confer absolute clarity, and some readers may well consider such extracts inconclusive. (Readers who may be concerned by the liability implications will find the subject more comprehensively addressed in Chapter 21.) The answers to these questions lie in BS 5750 itself which is normally the system Standard (and there are no other published criteria so comprehensive or so respected apart from a similar BS [5882] for specialized applications which is described in Chapter 3).

Having seen some of the interpretative difficulties of quality assurance we can now, introduce a reasonable working basis for understanding the real meaning, and, just as importantly, a meaning which is accepted by the principal certification bodies in the field. 'Quality Assurance' really means 'assurance that the process is properly managed'. Terms which have emerged as strong contenders for clarity and real meaning are 'Quality Management', and in its application, 'Quality System'. If we stay with these terms as defined in the BSs (and this book attempts to), we can perhaps start to make some sense of the subject.

Quality management is defined in BS 4778 Clause 3.5 as

> That aspect of the overall management function that determines and implements the quality policy

... and notes

> 1 The attainment of desired quality requires the commitment and participation of all members of the organization whereas the responsibility for quality management belongs to top management.
> 2 Quality management includes strategic planning, allocation of resources and other systematic activities for quality such as quality planning, operations and evaluations.

This definition, whilst wholly admirable and apt, might perhaps be more easily interpreted by the designer as

> Management of the process required (the process being the commission undertaken for the client) in order to achieve the quality promised to the client.

Let us conclude these definitions by having another look at 'quality'. For the designer it means

> The service undertaken by the designer (*supplier*) in the agreement for services (*the contract*) to his client (*the purchaser*).

(words in parenthesis are those used in BS 5750, a first introduction to readers of the sometimes difficult terminology of the BS for the designer). Thus, by implementing a quality system the designer is not committed to providing the quality promised; but he is committed to providing a level and quality of management appropriate to the services he undertakes. The definitions discussed above may be used with confidence by those who intend to adopt or have already adopted 'Quality Assurance'.

Are there any aspects of practice which should not come within a quality system? BS 5750 will be discussed in detail later, but it might be useful to note here that its main thrust is towards the management processes which influence the product or service provided by the firm. While the firm is free to include any aspect of its activities within its quality system (provided all proper aspects of BS 5750 are addressed), if the system is to be approved by a certification body, that body may be reluctant to assess aspects which fall outside the BS.

The principal aspects which certification bodies may not wish to assess are the organization's methods of controlling the cost of the product, and the time required to produce it. Although it is possible to identify a place in the BS which covers these aspects, the certification body may have to be persuaded. Designers may well consider time and cost to be critical aspects of quality, and should therefore give careful consideration to their inclusion. These aspects are developed in Chapters 5–16.

There are other areas for which inclusion in a quality system is more debatable. In broad terms, they comprise the organization's 'housekeeping' operations. They might include maintenance of the office, health and safety, fire precautions, and control of financial matters (not to be confused with cost control of the product, which was discussed above). Few of these are typically included in systems, but there is no reason why they should not be, if they are considered sufficiently important. This matter is developed in Chapter 23.

To Summarize

Definitions are given of

Quality

What is quality assurance?

- BS 4778 definition.
- Its meaning in practice to the building designer.
- 'Quality' is not necessarily synonymous with 'superiority'.

Assurance

- The dictionary definition.

Quality Assurance

- BS 4778 definition.
- Its meaning in practice to the building designer.
- 'Quality Assurance' is not 'an assurance of quality' or a guarantee of performance.

Quality management

- BS 4778 definition.

Quality system

Aspects of practice which may not be covered by a quality system.

2 The History of Quality Assurance

Although the term 'Quality Assurance' may not have been recognized, and certainly not in its current context, until after the Second World War, its essential features have existed wherever the provision of goods or services have been provided in exchange for a consideration (usually money). Quality assurance is a response to demand; by the purchaser of the goods demanding some assurance from the supplier that quality would comply with the purchaser's requirements; and by the 'supplier' who would attempt to secure competitive advantage for his product by 'assuring' a 'superior', or a 'best' quality.

Quality assurance as we know it today first appeared formally in a commercial situation. We can be more certain about the period when it began to acquire some significance. The start of the industrial revolution in the eighteenth century was the start of a process which would change the whole world more fundamentally and speedily than at any time before or since. Even if the principles of trading did not change, emphasis and practice certainly did. Purchasers soon began to demand certainty in quality as never before. Reliance on quality was an important factor in either making spectacular fortunes, or suffering spectacular losses. Correspondingly, suppliers were forced to seek what protection they could against redress sought by suppliers for alleged faulty goods and services. Thus were born comprehensive specifications and contract conditions in a harsh commercial climate, with no mercy given or expected. From this often crude and elementary process emerged the principles of quality systems as they are understood today. Manufacturers began to use standard specifications for similar products to replace the many differing standards demanded of or offered by individual suppliers. Purchasers had more confidence in products complying with industry-developed standards. The next stage was the development of independence in the bodies making the standards; in the standards making bodies; a development leading to the creation of the British Standards Institution (BSI) at the turn of the century.

Purchasers were, however, demanding more than industry-wide

specifications. They wanted some means of assurance that the specified product would, when completed, comply with the specification. The result was inspection and testing. Inspection and testing took many forms. Large purchasers (central and local government, water boards, railway companies) had sufficient resources to establish their own testing houses. Smaller purchasers initially had to rely on testing carried out by their suppliers. This somewhat incestuous process was overtaken to some extent by the growth of independent testing bodies, many of which still exist today, perform excellent services, and have high reputations. Here again, we see the seeds of BSI, this time as a certification body when BSI introduced kitemarking in 1926.

Government was determined that society would be better served by the establishment of an independent agency representing purchaser and supplier and financially underwritten by the government. This would be a certifying as well as a standards making body and would benefit all purchasers and suppliers, and would not just be for the protection of the government as a purchaser. Here was the final step in the creation of BSI which received its royal charter in 1929.

The end of the Industrial Revolution saw considerable stability in the control of manufacturing standards, comprehensive contracts, a legal system equipped to handle and pronounce on the most complex of disputes, industry-wide standard specifications, independent testing and certification of products.

Two initiatives in particular should be mentioned as having a seminal influence in the development of quality systems. Some of the earliest examples of QA are to be found in insurance, predating even the Industrial Revolution. Underwriting shipping risks has a long history. The Lloyd's market through which underwriting bodies insured shipping loss, undertook considerable risks. They demanded in return some assurance both as to the safety of the ships themselves and their cargoes. Thus the now well-known Lloyd's shipping standards and inspections were founded.

It was, however, in the defence industry where quality assurance was developed in the form which is recognized today; a new element in the development of quality systems. Security of the state has always been an important and sensitive area. Its meeting point with the commercial world which largely provides its hardware was consequently seen by the government as needing quality controls at least as stringent as those found elsewhere in commerce. The development of nuclear power added a new dimension to such sensitivity. On both sides of the Atlantic, governments, until recent years, relied on final testing by them of products they purchased from industry as the only means of assuring the quality they demanded. Consistently high failure rates and poor delivery times persisted. No doubt in the somewhat commercially artificial climate of defence the economic effects of such high failure rates could perhaps be tolerated. In other areas of commerce the purchaser might have to suffer a poor quality of product which the defence sector would not tolerate.

Governments began to research the reasons for such high failure rates, with the object of improving the performance image of industry as a whole, whatever its markets. It was found that final inspection of a product by the purchaser was largely the detection of failure; not the best way to achieve the assurance of quality. Moreover, the often heavy hand of the purchaser's inspector provided some opportunity for the supplier to abdicate responsibility for quality control.

Two important findings emerged. Firstly, there was need to control quality over the whole production chain; testing only the end-product was too late. Secondly, key personnel in the production chain must accept and be committed to maintenance by them of quality in the parts of the process they produced. This needed to be complemented by inspection within the purchaser's organization (now known in QA terminology as 'verification') as the prime means of assuring quality, whether or not the purchaser himself also inspected.

Rules and standards, initially for the defence industry, were written in attempts to establish quality system practices based on these findings. These developments reached the wider commercial world in 1971 when the Confederation of British Industries asked BSI to investigate whether the NATO quality control system requirements AQAP 1 and AQAP 2 could beneficially be adapted for industry-wide use. Was it feasible, it was asked, to write a national Standard which could be applied to any commercial production process?

In 1972 the first document was published by BSI in response to this initiative. It was BS 4891: 1972 A guide to quality assurance. While it was then, as it is now, correct to title it a 'guide', historical perspective enables us to see it more as a consultative than a guidance document. Industry response would determine the course QA was to take. Its foreword is prophetic

> It is hoped that this document will contribute toward satisfying some of the needs of industry and commerce in general. Although it has been issued as a guide and not in the form of a specification with mandatory requirements, it should assist companies or organisations engaged in developing their own capabilities to meet contractual requirements ...

BS 4891 put forward many features a quality system should have, including design/specification control, review, communication and documentation and, perhaps most significantly, the need for sound management objectives.

With the benefit of hindsight we can forgive the obvious uncertainty in the foreword to the 1972 BS in the interchangeability of the terms *Quality Assurance* and *Quality Control*. It is clear, however, and this is important to an understanding of the current nature of Quality Assurance, that BS 4891 emphasized QA as a means of managing the quality processes. It was not a vehicle through which the supplier could offer any guarantee, warranty or assurance that his product would satisfy the quality specified. There were probably two reasons why this was so. Firstly, the development of a national Standard with which every producer could comply, clearly could not be a universal quality warranty applicable to specific products, and secondly there were (and still are) specific product conformance Standards which satisfy such needs. There is no doubt that BS 4891 was the precursor of BS 5750. Parts 1–3 of 5750 comprise the requirements developed from the recommendations in BS 4891, whilst Parts 0 and 4 of BS 5750 comprise the guidance.

It can be seen from what was stated earlier why the origins of the current quality Standards are industry and particularly product-based. Even though BS 5750: Part 1 embraces *design/development, production, installation and servicing*, it was based on the presumption that all of its components were to be found in a single supplier organization with or without subcontractors. It was not then recognized that the supplying sources might in fact be provided by separate organizations offering 'design' and 'production', respectively. This may explain the quite unnecessary struggle by some organizations to apply the 'production installation and servicing' clauses of the BS to the design process.

Those who provide consultancy type services, and in particular those who provide building design services (for whom this book is written), came comparatively recently to the quality system scene. In the absence of any rules of their own, they have had to adopt the rules developed from two centuries of experience and development by the manufacturing sector. They will find, however, that even though the rules have been developed from a culture they may consider to be alien, the clauses of BS 5750 Part 1 are, when the production aspects have been discounted, surprisingly relevant to the design process, as indeed organizations as diverse as British Rail and the legal profession have discovered.

What has caused building design to be drawn into quality management? Historically, formal quality management systems were not as appropriate to designers as to industry, which since the start of the Industrial Revolution has faced the challenges of management in the complexities of production, the problems of subcontracting, commercial pressures, and many other factors. In contrast, design was produced by relatively small organizations where management control could be exercised by a few people. There was also a cultural divide (perhaps more assumed than real) where the 'learned' professions sought and protected their detachment from commercial activity. This was fostered by the universities who did not see their aims as being primarily vocational.

Over the last two generations such differences have been substantially reduced, even though they still exist. There are many reasons, including the widening of educational opportunity, the increase in consumer knowledge and power and government determination to remove the privileges of cartel maintained conditions of engagement and fees.

Against this background, public and private sector clients have increasingly demanded that the professions should accept the same quality system disciplines that they have accepted, either as suppliers themselves, or as prime developers of quality management. Correspondingly, the professions have found that they are operating in a commercial world and have accepted that they are susceptible to the same pressures as any other supplier of goods or services. Their own management now has to respond to the complexities of modern building design, where the management of quality can no longer remain in the hands of a single principal. It may not be so surprising that so much of BS 5750 is appropriate to design.

Quality assurance was directly introduced to the building designer following the Government White Paper of 1982 (see Chapter 4). Government gave direct stimulus to industry to practise quality management by asking the Property Services Agency (PSA), as its then major purchaser, to encourage suppliers to adopt QA. PSA's multi-professional building design arm, its Design Standards Office, set an example by adopting a BS 5750 quality system and obtaining a certificate of approval from an external certification body.

What lies in the future? It is interesting, and perhaps more a social comment, that until recently the development of quality assurance has concentrated totally on the direct quality related processes; the product itself and not the people who produce it. This phenomenon can be explained by going back to the Industrial Revolution roots of quality systems. Operatives' working conditions were regarded as the minimum necessary to produce the article; labour was cheap and expendable. Development of quality systems may not have kept pace with social change. A workforce's environment is increasingly seen as an important influence on the quality of what is produced. The Japanese have certainly recognized this as an important constituent of quality. It is strange, for example, that the current major forces in the drive for quality, quality management and health and safety at work, are seen as separate forces. Perhaps we should look to 'Total Quality Management' (see Chapter 23) as the next development?

To Summarize

Principles of QA first emerged in the industrial revolution as a demand for quality control in manufactured products. Formal recognition of QA arose after the second world war as a need to strengthen quality management in the defence industry. Government played a large part in the development of QA, including definitions and transfer of responsibility for inspection and testing from supplier to purchaser. Government encouraged recognition of BSI as the national standards making and certification body. BS 5750 developed from BS 4891 as the prime quality systems standard. In 1982 the government encouraged acceptance of BS 5750 as the national quality systems standard, and the growth of independent certification bodies. Demand in recent years for QA, and for certification, has increased in the service and design sections of industry.

3 The British Standards

Introduction

Unless the designer intends to write his own standards for the quality management of his organization, which would be a remarkable example of re-inventing the wheel, he will need to make the acquaintance of most, if not all, of the British Standards written on the subject. The purpose of this chapter is to make as painless as possible an introduction to Standards which at first sight do not relate easily to each other, much of whose content is written in language originally intended for manufacturing processes. This chapter attempts to draw attention to those parts of the Standards which are essential reading, and where necessary interprets from manufacturing into the language of the designer.

With increasing trade between countries there have been corresponding attempts to adopt common Standards. Typically Standards developed in one country have been accepted (or not) by other countries. In recent years there has been considerable progress at government level either in the making of existing Standards applicable internationally or in producing new internationally sponsored Standards.

As will readily be appreciated, this process is far from complete; even within the European Community with a considerable history of trading between member countries, much remains to be achieved in the agreement and production of common Standards. At this transitional time we find a sometimes confusing mixture of local and international Standards, each with their own numbering and classification systems. The Standards relating to quality are caught within this process, where the picture is no less difficult to understand. This is particularly acute for the 'family' of quality Standards reviewed in this chapter, where consistency between them is sometimes elusive. Most of the present quality Standards were initiated by the UK and have been accepted internationally, but the latest are being written under the umbrella of full international co-operation.

The process of internationalization of Standards was started by the International Organization for Standardization (ISO), and the series of

Quality Standards referred to in this book were adopted by the European Committee for Standardization (CEN). CEN developed a Standards system for member states of the European Community and the European Free Trade Area (Austria, Belgium, Denmark, Finland, France, Germany, Greece, Ireland, Italy, The Netherlands, Norway, Portugal, Spain, Sweden, Switzerland, and the UK). The secretariat is in Brussels. Some of these Standards have also been adopted by other western industrial nations.

Although the ISO and CEN equivalents will be of interest to those trading outside the UK, for ease of use and simplicity, the Standards are referred to in this book by their BS equivalents. Figure 3.1 shows the ISO and CEN equivalents to the British Standards, and Fig. 3.2 shows the relationship of Quality Standards to each other.

Categories of Standards

The Quality Standards which are described in detail later may be seen as coming under three categories (categorization used here is not necessarily as expressed in the Standards themselves).

Quality System Standards

Quality System Standards are those which are adopted by organizations who seek to demonstrate that they have a quality system which conforms with a given Standard. While British Standards have only the status of recommendations, once they are adopted they become effectively the rules for compliance. Of course, an organization may wish simply to state that it has complied with rules of its own making, but it is more usual that a certification body will undertake to assess an organization for system conformance with one or more of the British Standards. The Standard normally used in the UK for building design activities is BS 5750: Part 1, around which this book has been written. Part 2 and 3 are also conformance Standards, but are not normally relevant to the design process, and are not described.

BS 5882 is another Quality System Standard which, while intended for organizations who manufacture and install in the nuclear field, may be a requirement by nuclear clients for the design process only. It may be thought that building commissions which are not critical to nuclear processes would be better assessed under BS 5750: Part 1 (which is acknowledged in the foreword to BS 5882).

Guidance Standards

Guidance Standards are intended to help organizations who write and implement quality systems which comply with Quality System Standards.

Fig. 3.1 Standards, alternative sources and equivalents

British Standards Institution (BSI)		Institutional Organization for Standardization (ISO)		European Committee for Standardization (CEN)	
No.	Title	No.	Title	No.	Title
BS 4778: Part 1: 1987	British Standard Quality vocabulary Part 1: International terms	ISO 8402 — 1986	Quality Vocabulary	—	—
BS 4778: Part 2: 1979	British Standard Quality vocabulary Part 2: National terms	—	—	—	—
BS 5750: Part 0: Section 0.1: 1987	British Standard Quality systems Part 0: Principal concepts and applications Section 0.1 Guide to selection and use	ISO 9000 — 1987	Quality management and Quality Assurance Standards: guidelines for selection and use	EN 29000 — 1987	Quality management and Quality Assurance Standards: guidelines for selection and use
BS 5750: Part 0: Section 0.2: 1987	British Standard Quality systems Part 0 Principal concepts and applications Section 0.2 Guide to quality management and quality system elements	ISO 9004 — 1987	Quality management and quality system Elements: guidelines	EN 29004 — 1987	Quality management and quality system Elements: guidelines
BS 5750: Part 1: 1987	British Standard Quality systems Part 1 Specification for design/development, production, installation and servicing	ISO 9001 — 1987	Quality systems — model for Quality Assurance in design/ development, production, installation and servicing	EN 29001 — 1987	Quality systems — model for Quality Assurance in design/ development, production installation and servicing

BS 5750: Part 2: 1987	Part 2: Specification for production and installation	ISO 9002 — 1987	Model for Quality Assurance, production and installation	EN 29002 — 1987
BS 5750: Part 3: 1987	Part 3: Specification for final inspection and test	ISO 9003 — 1987	Model for Quality Assurance in final inspection and test	EN 29003 — 1987
BS 5750: Part 4: 1990	Quality systems Part 4 Guide to the use of BS 5750: Part 1: Specification for design/development, production, installation and servicing, Part 2: Specification for production and installation and Part 3: Specification for final inspection and test	ISO 9004 — 1987	Guide to quality management and Quality system elements	EN 29005
BS 5750: Part 8: 1991	Quality systems Part 8. Guide to Quality management and Quality system elements for service			
ISO 9004	Quality management and Quality system elements, Part 2: Guidelines for services			
BS 5882	Specification for a Total Quality Assurance programme for nuclear installations			
BS 7000	Guide to managing product design			
BS 7229: 1989	British Standards Guide to Quality systems auditing		*Draft* Generic guidelines for auditing quality systems	

Fig. 3.2 Relationship between System, Guidance and Vocabulary Standards

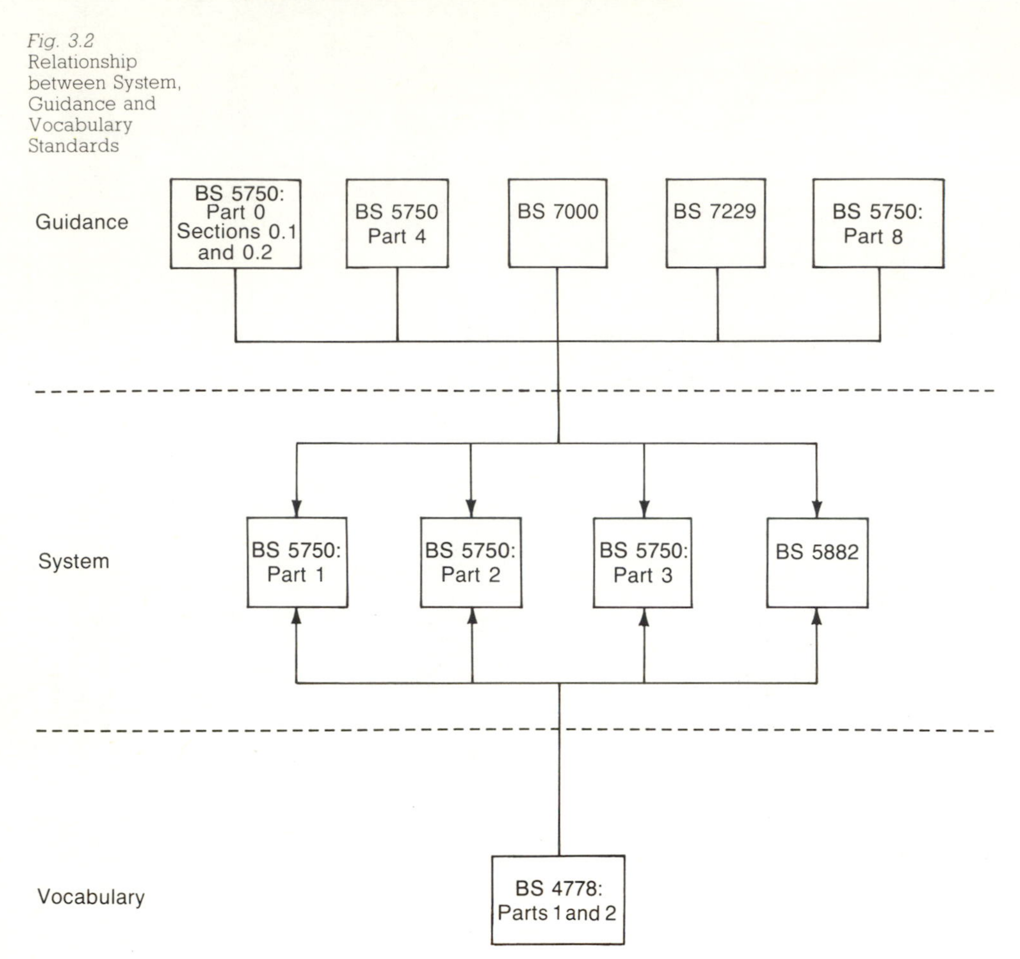

Organizations are not obliged to use them, and they are not intended to supersede or vary Quality System Standards. Indeed, the foreword to BS 5750: Part 4 neatly summarizes the relationship between Quality System and Guidance Standards

> This part of BS 5750 does not add any requirements to those in BS 5750: Parts 1, 2, and 3. Care should be taken that the guidance given, which may suggest various methods of providing assurance, is not used in place of, or as additional requirements to that given in BS 5750: Parts 1, 2, and 3.

It is as well to note, however, that the certification bodies normally base their assessment processes on the Guidance and Quality Vocabulary Standards (below), so an organization which intends its quality system to be approved by a certification body may find it helpful to read and, where appropriate, to adopt the contents of these Standards in the construction and operation of its quality system.

Guidance Standards are:

BS 4891
BS 5750: Part 0, sections 0.1 and 0.2
BS 5750: Part 4
BS 7000
BS 7229

Quality Vocabulary Standards

Quality Vocabulary Standards define the key terms used in the System and Guidance Standards. Some terms have acquired considerable importance as will be described later in this chapter, and should be regarded as 'compliance terms'; BS 5750: Part 1 states (Clause 3: Definitions)

> For the purposes of this International Standard the definitions of ISO 8402 (BS 4778: Part 1) apply

BS 4778: Parts 1 and 2 are the Quality Vocabulary Standards (Part 2 will be replaced and a new Part 3 added after this book goes to press).

The Standards in Detail

There is no substitute for reading the Standards themselves; each reader must decide for himself which aspects of the Standards are appropriate to his organization. However, it is recognized that, as most of the Standards were developed for the manufacturing process, it may be helpful to describe the sections most appropriate to the design process.

System Standards

BS 5750: Part 1 (Specification for design/development, production, installation and servicing)

Chapter 7 contains guidance on the interpretation of BS 5750: Part 1. Clause headings only are given here to indicate their scope and coverage. (Clauses 1, 2 and 3 cover *scope, references and definitions* respectively.)

Clause	
4	Quality system requirements
4.1	Management responsibility
4.1.1	Quality policy
4.1.2	Organization
4.1.2.1	Responsibility and authority
4.1.2.2	Verification resources and personnel

4.1.3.3	Management representative
4.1.3	Management review
4.2	Quality system
4.3	Contract review
4.4	Design control
4.4.1	General
4.4.2	Design and development planning
4.4.2.1	Activity assignment
4.4.2.2	Organizational and technical interfaces
4.4.3	Design input
4.4.4	Design output
4.4.5	Design verification
4.4.6	Design changes
4.5	Document control
4.5.1	Document approval and issue
4.5.2	Document changes/modifications
4.6	Purchasing
4.6.1	General
4.6.2	Assessment of subcontractors
4.6.3	Purchasing data
4.6.4	Verification of purchased product
4.7	Purchaser supplied product
4.8	Product identification and traceabilty
4.9	Process control
4.9.1	General
4.9.2	Special processes
4.10	Inspection and testing
4.10.1	Receiving inspection and testing
4.10.2	In-process inspection and testing
4.10.3	Final inspection and testing
4.10.4	Inspection and test records
4.11	Inspection, measuring and test equipment
4.12	Inspection and test status
4.13	Control of nonconforming product
4.13.1	Nonconformity review and disposition
4.14	Corrective action
4.15	Handling, storage, packaging and delivery
4.15.1	General
4.15.2	Handling
4.15.3	Storage
4.15.4	Packaging
4.15.5	Delivery
4.16	Quality records
4.17	Internal quality audits
4.18	Training

4.19 Servicing
4.20 Statistical techniques

BS 5882: 'Specification for a total quality assurance programme for nuclear installations'

The scope and compliance for this BS is similar to that of BS 5750, so any attempt at interpreting into designer language would, to a great extent, be repetitious. Unlike BS 5750 it has its own definitions which are a further aid to the reader who seeks to understand it.

The emphasis is different because of the particular hazards of working with or producing nuclear materials, the need for high standards of design and inspection, and the need to reassure the public that these measures are maintained. Thus it will be found that a higher client involvement is stated and can be expected than with typical systems which comply with BS 5750. The client is likely to demand that a part or a whole of the quality system complies with BS 5882 and that he inspects with some rigour that the system is in place and is being implemented.

Guidance Standards

BS 4891: 1972 A guide to quality assurance

BS 4891 was described in Chapter 2. It is more a basis for the development of BS 5750 than a currently used Guidance Standard. It is of interest in providing historical perspective to the development of quality assurance.

BS 5750: Part 0: Section 0.1 (guide to selection and use)

This section of BS 5750 has two purposes: to state the objectives to be accomplished with regard to quality; and to give guidance on the Standards which may be used for quality management purposes.

The Standards are all reviewed in this book, but the following clauses are particularly worth reading in setting the quality scene before devising the formal quality system.

4 Principal Concepts

The three objectives that have to be accomplished

— Meet the purchaser's stated or implied needs
— Confidence within the organization that intended quality is being achieved
— Confidence to the purchaser that the intended quality is being achieved.

5 The characteristics of quality system situations

Contractual and noncontractual emphases. (The differences between a system which exists purely for the well-being of the organization, and a system which is offered or demanded as part of the contract between purchaser and supplier).

8.3 Demonstration and documentation
Some aspects of a quality system which have to be addressed.

8.5.1 Tailoring

This is a fairly important clause invoked by BS 5750 which permits the purchaser and supplier to agreement to add to or delete from Quality System Standards. This is the clause which will permit the designer to delete (i.e. exclude from his system) those clauses in BS 5750: Part 1 which refer solely to the manufacturing process. The clause states that such changes should be recorded in the contract. However, it is more likely that the supplier will record such changes in his Quality System Manual (see Chapter 8 for guidance on writing the Quality System Manual) and give the client a copy. It is also advisable to consult the certification body where appropriate, if changes are contemplated.

Guidance Standards: BS 5750: Part 0, Section 0.2 (Guide to quality management and quality system elements)
(See also commentary following, on Part 4)

This section develops the principles of Section 0.1 into the practice of detail in the development of a working system. It is too long and intricate a Standard for detailed review, but it is recommended reading. It may be helpful, however, to attempt to classify its content under the following headings:

1 General principles of quality management
- 0 Introduction
- 1 Scope
- 2 References
- 3 Definitions
- 5.1 Quality Loop

2 Specific advice on setting up aspects of the system which relate directly to clauses of BS 5750: Part 1
- 4 Management responsibility
- 5.2 Structure of Quality System
- 5.3 Documentation of the system
- 5.5 Review and evaluation of the Quality Management System
- 8 Quality in specification and design
- 9 Quality in procurement
- 13 Control and measuring of test equipment
- 14 Nonconformity
- 15 Corrective action
- 17 Quality documentation and records
- 18 Personnel
- 19 Product safety

It may be advisable to study BS 5750: Part 1 first to assess what is demanded; the above clauses will then assist in writing and implementing the system.

3 Specific advice on setting up aspects of the system, parts of which are not directly required by BS 5750: Part 1

6 Economics — quality related cost considerations
7 Quality in marketing

4 Specific advice on areas covered in more detail by other Standards

5.4 Auditing the quality system (BS 7229 seems more comprehensive in treatment of the same subject)

5 Areas that are largely unlikely to be relevant to design

11 Control of production
12 Product verification
16 Handling and post-production functions
20 Use of statistical methods

BS 5750: Part 4 (Parts 1, 2 and 3)

This Part should be seen as augmenting Section 0.2 (above).

BS 5750: Parts 0, 1, 2 and 3 were written in a language for, and emphasis on the manufacturing process. As a Guidance Standard, Part 4 was written partly in recognition of the fact that designers needed different language and emphasis, but principally as a valuable indication of how the specific requirements of the Quality System Standards might be addressed in developing quality documentation and practice. Part 4 refers directly to the paragraphs of Parts 1, 2 and 3. However, it addresses all aspects, and therefore contains some manufacturing-based material which is of no direct interest to the designer who is devising a system.

Its relationship to Section 0.2 is not entirely clear; the emphases and coverage differ somewhat. 0.2 is an International Standard and Part 4 is a British Standard. The introduction to Part 4 makes it clear that Part 4 is not a substitute for 0.2 *which has its own distinct relationship to Parts 1, 2 and 3*. The best advice, perhaps, is to read the corresponding sections of 0.2 and Part 4 together.

BS 7000 (Guide to managing product design)

The word 'product' in the title should not mislead the reader who might initially draw the conclusion that the Standard is directed more towards the design of manufactured articles. Although the scope of the Standard expressly limits its application to products, much of the context and expression are entirely relevant to the building design process.

The Standard comprises recommendations and checklists for the management of design, from corporate responsibilities to activities at the drawing board. It complements BS 5750: Part 1, particularly in the specific design management activities covered by the BS. The overall aim of BS 7000 is to improve 'product design'. Readers who are developing quality manuals, design related procedures and methods of compiling job quality plans will find this Standard contains much helpful advice, written in a practical and readable manner.

BS 7229 (British Standard Guide to quality systems auditing)

Auditing is an important part of ensuring that the quality system is properly maintained. This BS is an extremely useful and practical document, second only in importance to BS 5750: Part 1 in the design and application of a quality system. It describes the purposes of audit, the different kinds of audit, and how to audit, all in manageable detail. Following this BS should ensure both that the audit requirements of BS 5750 have been complied with, and that the audit procedures satisfy the certification body.

BS 5750: Part 8: 1991 Guide to quality management and quality system elements — guidelines for services

This Standard provides a basis for *quality systems* in respect of a wide range of services including *medical practice, catering, banking, building design and legal services*.

Features of particular note in the Standard are

- *Service brief*: this defines the customer's needs and forms a basis for the design of a service.
- *Design process*: this covers the process of designing a service which involves converting the brief into specifications for both the service and its delivery.
- *Design review*: this is a review of the design of a service against the service brief.

As will be seen, the expression *design process* is used to denote the process of designing the service. This may be confusing for designers who naturally think of the design process as the service they provide in producing the design.

As the potential scope of BS 5750: Part 8 is so wide, e.g. services ranging from catering to computing, it requires a considerable amount of interpretation. It is too early to say how well this Standard will work in practice and how widely it will be used. It should, however, at least provide a useful check list against which to examine procedures for services which are not covered by other standards, e.g. BS 5750: Part 1 which includes design.

Vocabulary Standards

BS 4778: Part 1 (Quality vocabulary Part 1. International terms)

This Standard is particularly important in defining the more important terms used in the compliance Standard, BS 5750: Part 1. Indeed, as stated earlier in this chapter, BS 5750: Part 1 invokes BS 4778: Part 1. Thus, the Standard should be seen as a compliance Standard for the terms it defines which are used in BS 5750: Part 1. Even for terms that it defines which are not used in BS 5750: Part 1, it is an important Standard, because they are all commonly accepted quality definitions, and many are used in other quality Standards.

Some of the terms in this Standard have already been given in full in Chapter 1. All the terms which are considered important in the operation of quality systems are given below, and a study of their definitions in the Standard is recommended reading (terms which do not relate directly to design related activities have been omitted).

3.1 Quality
3.4 Quality policy
3.5 Quality management
3.6 Quality assurance
3.7 Quality control
3.8 Quality system
3.9 Quality audit
3.11 Quality surveillance
3.12 Quality system review
3.13 Design review
3.14 Inspection
3.15 Traceability
3.20 Nonconformity
3.21 Defect
3.22 Specification

BS 4778: Part 2 (Quality vocabulary Part 2. National terms)

It might be assumed from the titles of the two parts of the Standard that Part 2 contains terms which are used only in UK, but otherwise the purpose and approach would be similar. Such an assumption would not be entirely accurate. Part 2 in fact attempts conceptual definition, with much of its content devoted to complex manufacturing based definitions, which are clearly not relevant to the building designer. Moreover, this Standard is not invoked by BS 5750: Part 1, so it might more usefully be seen as a Guidance Standard for the terms it defines which are relevant to design. The terms defined in this Standard which may usefully be studied in the development of a quality system for a design based operation are given below (terms which do not relate directly to design related activities have been omitted).

4.1	Concept of quality
5.1	Concept of quality assurance
6.1	Concept of specification
7.1	Design
8.1	Conformity
8.3	Accuracy
8.4	Precision
8.5	Proficiency
8.6	Timing
9.1	Compliance
11.1	Quality manual
11.3	Quality programme
11.9	Quality verification
11.15	Calibration
12.3	Quality related costs
12.5	Appraisal costs
12.6	Failure costs, internal (seen to be relevant, even though specific to manufacturing)
12.7	Failure costs, external (ditto)
15.1	Concept of grade
15.2	Operational requirements
15.3	Configuration
15.4	Target specification
17.1	Functional specification
18.1	Installation specification
18.2	Use specification
19.1	Concept of quality control

The subject is not, of course, a part of the scope of this book, but designers may find some of the definitions which relate to manufacturing useful when writing specifications for the parts of the construction process which are similar to the manufacturing process.

To Summarize

This chapter describes the principal national and international quality Standards relevant to building design.

Quality System Standards

BS 5750: Part 1, BS 5882 which lay down the requirements of quality systems which firms must comply with if they adopt either of these Standards.

Guidance Standards

BS 4891, BS 5750: Part 0, sections 0.1, 0.2, BS 5750: Parts 4 and 8, BS 7229, which offer interpretative and operational guidance intended to assist firms in the operation of systems based on BS 5750: Part 1.

Quality Vocabulary Standards

BS 4778: Parts 1 and 2 which give definitions of some of the terms used in the Guidance and Systems Standards.

Attention is drawn to the sections of these Standards which should be studied when compiling a quality system.

4 The Certification Bodies and NACCB

Historical Context

Certification has a long history, and can be found in many forms. Typically a purchaser of goods or services will require certification that commodities which have a high risk potential comply with given standards. Governments often insists on forms of certification, particularly for products which pose a potential danger to life or safety. The purchaser, particularly if he is from a large organization, may have developed his own certification processes. Sometimes the purchaser has demanded conformance with standards not written by him, and instead of carrying out his own checks has accepted or required certification by an independent body. British Standards Institution (BSI) has developed its services within this range of experience and has become a major certification body as well as the publisher of national Standards. A BSI kitemark gives assurance of conformance based on system assessment and product testing. Thus BSI acts as a certification body for product conformity. Where a producer wishes to use a kitemark on his product, BSI will either test or verify that testing was carried out in accordance with the appropriate product Standard. (If the product is found not to be within the required acceptance criteria, the producer cannot use that kitemark.)

Certification bodies undertake different kinds of work; either system application or system application and product conformity certification. For example, BSI and CARES (who test reinforcing steels against British Standards) certify both; Lloyd's Register Quality Assurance (LRQA) and Construction Quality Assurance (CQA) only certify system applications. The British Board of Agrément (BBA) tests against its own standards.

Government was aware of the vast and sometimes duplicated or conflicting practice in British industry in its attitudes to standards and certification, and decided that some rationalization was necessary as one of the means of improving Britain's quality image in overseas markets.

In July 1982 the Department of Trade published the White Paper 'Standards, Quality, and International Competitiveness'. This document was the base from which accreditation of certification bodies has been developed.

Its thrust was largely devoted to raising the status, quality and awareness of nationally produced goods and services. It was made clear that the Government wished to promote BSI as the major focus and organization in the production of Standards. A Draft Memorandum of Understanding between the Government and BSI was included in the Paper, and subsequently enacted. Considerable emphasis was also given to the importance of QA as one of the means of strengthening the international competitiveness of British goods. The White Paper encouraged the development of independent assessment (certification bodies) and it encouraged the adoption of BS 5750 as the principal basis of approach and discipline to QA. Thus two important steps were recommended; promotion of the importance of the independent certification body, and the implied assumption that certification would be for conformance of quality management systems with BS 5750. The Paper took a third important step. It recognized that to have the authority and status necessary for home and overseas markets, certification bodies would need to operate within unified and government approved criteria. They would have to demonstrate that they had applied given minimum standards in the certifying process. The Paper advocated a 'national accreditation mark' which would be awarded to approved certification bodies, and a central agency which would lay down standards to be attained by them, and which it would license to award the mark. Thus was born the National Accreditation Council for Certification Bodies (NACCB).

The National Accreditation Council for Certification Bodies

The White Paper recommended unified arrangements for the accreditation of certification schemes, so that certification would have the credibility expected of a Government sponsored process. The Paper explored two ways of overseeing such a unified arrangement, either by the Government itself or by an independent body. If it were to be an independent body, BSI was a strong contender because of its considerable experience in the writing of standards and operation of certification schemes. There was, however, a rider; that if BSI were chosen there must be organizational separation from its commercial certifying operation, similar to the separation which already existed between its standards writing function and its certifying function.

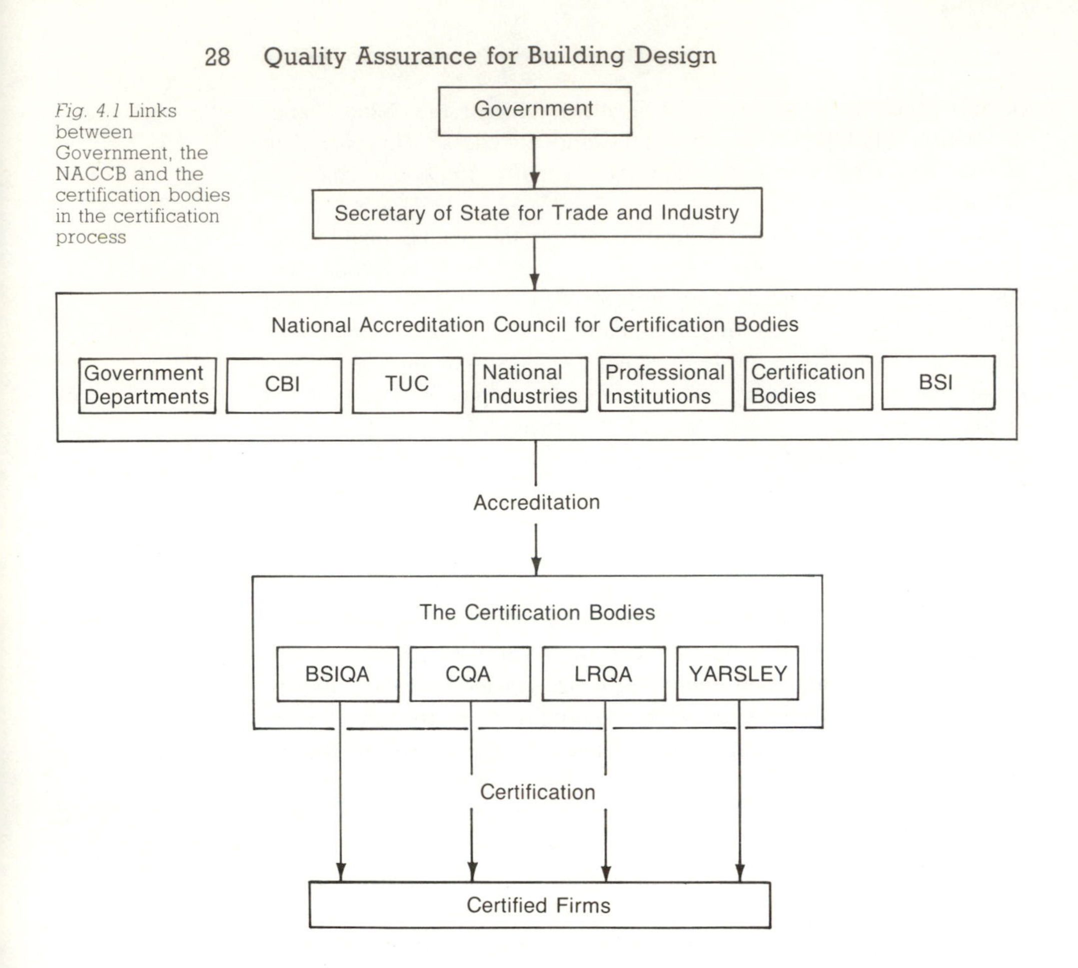

Fig. 4.1 Links between Government, the NACCB and the certification bodies in the certification process

In fact this is what has happened. The Council of NACCB operates under the terms of a Memorandum of Understanding between the DTI and the BSI. Independently of both, it makes recommendations to the Secretary of State for Trade and Industry. It is widely representative of QA interests, comprising Government departments, the CBI, professional institutions, certification bodies, and BSI itself. Its other accreditation functions deal with product conformity, product approval, and certification of personnel, but we are concerned here principally with its function towards certification of quality systems to BS 5750.

NACCB invites certification bodies to apply for accreditation for the categories of certification they offer. It then assesses them to ascertain whether or not they conform to its regulations and criteria, which are that accredited certification bodies shall be impartial, competent, free of conflicting interest, and staffed by competent personnel, their management being such that no one interest predominates. If they conform, they are permitted to market themselves under the accreditation mark; organizations whom they approve or register are also permitted to use this mark in conjunction with the mark of the certification body.

In laying down the above, NACCB have published several regulatory documents, including:

- NACCB rules of procedure.
- Prospectus (which describes essentially how NACCB processes applications for accreditation).
- Regulations governing the accreditation of certification bodies.
- Criteria of competence for accredited bodies (this and the regulations for accreditation above should be read by those who are interested in the criteria NACCB apply).
- Regulations governing the use of the accreditation mark (seen in Fig. 4.2, incorporated in certification marks by certification bodies who have been accredited by NACCB).
- Directory of accredited certification bodies.

It may not be necessary for the reader to enquire too deeply into the regulations required of certification bodies by NACCB who are aware of the areas of activity which require rigour, and assess the certification body accordingly. The main thrust of NACCB criteria and its assessment is to ensure that certificates are justified, products must comply with product Standards and quality systems to BS 5750. It seems clear from experience that the NACCB regulations and their application have won NACCB a sound reputation. Therefore it should be safe to assume that certification bodies who have been accredited have demonstrated appropriate competence and rigour.

It should, however, be remembered that NACCB confine their accreditation to the areas of expertise offered by the certification body, and that these areas do not appear on the accreditation mark. Those seeking certification should therefore ensure that potential certification bodies have been accredited by NACCB in the areas appropriate to the functions to be assessed. This information can be found in the Directory of Accredited Certification Bodies, available from NACCB or in the DTI QA Register (HMSO). Firstly the certification body should be seen to be accredited under category 1 'Certification of Quality Management Systems (to BS5750)'. Then, and perhaps more importantly, the 'accredited scope' should be examined. NACCB issues each body with a schedule covering the scope, and thus the identifying words which the body should use in marketing itself and must use on the registration or approval certificate it sends to its client. (For example, the relevant part of BSI Quality Assurance accredited scope under category 1 is 'multi-disciplinary building, civil and structural engineering design'.) Some certification bodies are now venturing into specialized sectors which are new to them. While they will be able to claim, say, the right to quote

Fig. 4.2 Examples of the certification marks for four certification bodies who have been accredited by NACCB to certify aspects of building design. Organizations approved or registered by these bodies may use the marks on their stationery and marketing material

BSI Quality Assurance

Construction Quality Assurance

Lloyd's Register Quality Assurance Ltd

ISO 9000 series of standards is the international equivalent of the BS 5750 series (see Chapter 3)

Yarsley Quality Assured Firms Ltd

Yarsley also license certification marks for NACCB schemes showing the alternative European and International Standards, and equivalent marks without the NACCB mark for schemes not covered by accreditation

the generic term 'architecture', (as in the case of LRQA), the client with a particular wish to market specifically 'landscape architecture' may wish to ensure that his certification body has NACCB accredited scope for this activity.

The Certification Bodies

Function

In the White Paper, Government stated that 'the standard for quality systems BS 5750 is the heart of the modern approach to quality assurance'. This statement has been adopted in practice and, unless there are special circumstances (e.g. nuclear industry requirement for compliance with BS 5882), the great majority of quality systems are now measured against BS 5750. Thus it is the certification body's prime function to certify that an organization's quality system complies with this BS (Part 1 for design organizations). Certification bodies aspiring to accreditation by NACCB will offer their clients the right to use the accreditation mark as well as their own mark. (See Fig. 4.2 for the marks used by the certification bodies.)

Constitution

Whether the certification body is a nonprofit-making arm of some institution, or an independent profit-making company normally has little bearing on its suitability for the organization seeking certification. However, if it is a subsidiary of another body the sectors covered by the certification body may be influenced to some extent by the principal body.

Sector Coverage

Certification is relatively new to building design, but a picture is emerging of the major certification bodies' interest. The names of the four bodies and their certification marks which currently cover this field are given in Fig. 4.2.

Any certification body can offer to certify any process. Care must be taken to ensure that it has a track record in the sector required or, if it has no experience, to enquire about its plans for acquiring the necessary expertise. It may not at the time of its appointment be licensed by NACCB to certify in the appropriate category, but could have considerable experience in a closely allied discipline where it is accredited. It may be to its and its client's mutual advantage to work together in developing an assessment process which, by the time it is ready to approve the client's quality system, will coincide with its demonstration to NACCB that it is qualified to certify the system.

Some indication of the certification body's suitability would be its experience in design-related fields; experience in product manufacture or construction activity alone would not be a good grounding. The process is relatively new; many certification bodies are still learning on the backs of practitioners, and this must be expected.

The Association of Certification Bodies

There is an Association of Certification Bodies. It is a non-profit-making organization concerned with furthering the interests of independent certification. Membership comprises third-party certification bodies (including the four bodies whose certification marks are reproduced above) who issue certificates of conformance to national and international Standards. The Association provides a forum for discussion and formulation of policy on matters of common interest. It is represented on NACCB.

Qualification of Assessment Teams

NACCB demand only that 'certification staff shall be competent for the function they undertake'. This is a fairly imprecise requirement, and certification bodies have interpreted it in several ways. The adequacy of an assessor derives from his formal qualifications and experience. Most certification bodies will field a team led by a 'lead assessor', someone who has been approved by the Assessors' Registration Board as having achieved the qualifications and experience required by the National Registration Scheme for Assessors of Quality Systems. It is as important to the firm to be assured that the team comprises assessors who are experienced in the sector. If such experience is lacking, an assessor unfamiliar with an industry may miss points essential to an effective quality system for the client. Furthermore, client time can be wasted by having to answer questions about the nature of the business or operation. It must be expected, however, in the current phase of great expansion, that a part of the assessment team will be trainees (whom the certification body may not charge to the client).

Subcontracting

Some certification bodies subcontract assessment work, others say they never do. Provided subcontractors are trained, experienced, fully briefed and vetted by the employing certification body, there seems no disadvantage to the firm. However, it has been observed that considerable use of subcontractors may not produce the harmony expected of a team comprising employees of the certification body who have worked with each other for some time. The firm may have concerns about possible breaches of confidentiality when a subcontractor assessor is also a potential competitor. Certification bodies are normally sensitive about this possibility, but firms should nevertheless bear it in mind.

The Certification Body's Assessment Process

The following description of the way the process operates is an amalgam of the procedures adopted by the four certification bodies currently operating certification schemes for building design organizations. The individual body's procedures may be slightly different and should be confirmed when the body is approached; what is given here is therefore a guide to what might be expected.

Initial Meeting

It is normal for there to be an initial meeting between client and certification body to explore the matters covered by this chapter. Some certification bodies will charge for this meeting.

Document Review

The first important stage in the process will be document review. This is a demonstration to the certification body that the quality system documents adequately reflect the operations to be covered, and that they comply with the BS. Certification bodies are generally reluctant to approve these documents in advance of assessment, because they will wish to satisfy themselves that the documents can operate adequately in the context of the operation being assessed. Certification bodies will stress that they cannot act in a consultancy role in the development of documentation; their role is only to certify what is put before them.

Assessment of System Operation

Assessment on site follows. For the building designer, this does not necessarily mean a building site, it means the place where the operation is carried out. This process is normally known as 'an audit'. A typical programme for an audit visit is an opening meeting in which the lead auditor (assessor) will state his intentions, followed by interviews with job teams and other staff to ascertain that they are following the system documentation previously reviewed. The audit team can be expected to issue formal non-conformance notes at the time of audit, and by the end of the visit they will inform the client whether or not his system can be approved, or, if not, the areas which require rectification. If corrective action is required a further visit or visits may be arranged and the process repeated until the certification body is satisfied that the system and its application comply with the British Standard.

All certification bodies will require that they make regular surveillance visits to ensure that the system and its application continue to conform to the BS. Some bodies make a complete re-assessment of the firm's quality system every three years; others will treat the system as continuous, and only make surveillance visits. Assessment can treat the firm as a whole or, if it has several offices, certification can be given to each office.

Registration or Approval

When registration (or approval, slightly different descriptions have been adopted by the certification bodies) is given by the certification body, a certificate will be issued to the firm which will show the scope covered, and whether or not the NACCB accreditation mark can be used.

The Certification Body's Schedules and Quality System Supplements

Some certification bodies have developed schedules or quality system supplements. There has been some confusion about the status of such schedules, particularly as they use terminology different from that used in BS 5750. LRQA state that their quality system supplements (QSS) are 'to give guidance in interpretation of assessment terminology for specified types of system. The QSS does not add requirements to the Standard (BS 5750). It does not prevent a company complying with the Standard in an alternative way (except where the guidance is nationally agreed*). It is not an assessment Standard. Noncompliances are only issued against the Standard'. Clearly, LRQA intend the QSS only to amplify and not in any way to supersede the BS, or to add further requirements.

BSI Quality Assurance requires compliances with the BS and their relevant quality assessment schedule (QAS). Their QAS for architectural services, like the LRQA QSS, is 'complimentary [*sic*] to BS 5750 and clarifies its requirements in relation to architectural services'. However,

* 'nationally agreed' may arise where statute or some government regulation demands adoption of a QSS. A similar situation may also arise either where a professional institute agrees a QSS with a certification body, and requires its members to comply, in the operation of their quality systems, or the market comes to expect that certificated firms' systems will comply with a particular interpretation of the BS. There are not, as yet, any such demands or requirements for building design firms.

its requirements are much more specific than those made by the BS. The following brief and incomplete summary of the QAS illustrates this point:

Records
A defined minimum retention period for quality system records is demanded.

Organization
A QA Manager and a deputy must be appointed.

Quality planning
Detailed requirements are set down in the control of documentation, including quality plans.

Project control
Detailed task requirements are set down, including project control by a 'registered architect'.

Information system
Minimum documentation to be held is required.

Training
Control of personnel skills, records, minimum criteria for other design disciplines engaged are demanded.

Complaints
Constituents for a procedure for dealing with complaints is prescribed.

There is now a general move away from these complementary documents by the certification bodies. Some organizations have, or are preparing, guidance documents putting BS 5750 lists into the context of the activities of the organizations they certify. However, the above summary has been included to remind firms that they should enquire into the precise intentions of their proposed certification body in regard to any QAS or QSS guidance document it may wish to apply in its assessment process. While the above extract may be a valid interpretation of the BS for most aspects of architectural practice, the firm should nevertheless ascertain that its own system can, and should, comply with any QAS, DSS or guidance documents put forward by certification bodies.

Publicity

Most certification bodies publicize themselves and sometimes the organizations they certify. If the firm intends to use certification as part of its marketing, it should enquire about the extent of the body's publicity machine.

In general it is difficult to locate a single source of organizations and their activities which have been approved by certification bodies. NACCB, as stated earlier, maintains a list of NACCB approved firms, but there are other bodies who do not come within NACCB approval. The DTI maintains a large and expensive register of all certified firms, listing them under the products or services they supply. The certification bodies themselves maintain lists of the firms they have approved.

Contracts and Charges

All certification bodies require their clients to enter into a formal contract. Prospective clients should ask to see the conditions at an early stage in negotiations.

The certification bodies' basis of charging varies considerably, and each produces a current tariff. Some charge by the day, some by the hour, and all will give a firm quotation for the process. Readers should be cautioned, however, that if return visits are required through nonconformance, such visits may be charged as extras. This is a good reason why assessment should not be premature; the importance of a firm's satisfaction that it is ready for external assessment (e.g. via an internal audit) cannot be overemphasized.

To Summarize

This chapter describes:

- The background to and history of certification and the particular part played by BSI.
- Government initiative through the 1982 White Paper which increased the status of QA by commending BS 5750 as the nationally agreed Standard for quality systems, encouraged the expansion of independent certification bodies and proposed national accreditation.
- The purpose and function of the National Accreditation Council for Certification Bodies (NACCB); how government control over the quality and standards of the certification bodies is maintained by their accreditation by NACCB.
- The purpose and function of the certification bodies; to certify that firms have complied with given Standards (normally BS 5750: Part 1).
- The certification bodies' assessment policies in complying with the above; advice on how the bodies' schedules and supplements should be approached; the bodies' contracts and charges.

5 Is QA Applicable to Your Organization?

Where QA is Applicable

Usually, QA is adopted for one or more of the following four reasons:

1. It offers consistency in the quality of service through an ordered approach to quality management thus improving risk management and minimizing negligence.
2. Clients are increasingly demanding evidence that their consultants have formalized quality systems.
3. The potential of a recognized quality system for improving market image.
4. It is fashionable.

The above are given in an approximate order of importance. Let us now take each in reverse order and examine their credentials.

Fashion

We suspect that not many organizations would invest in quality assurance for this reason alone. Nevertheless designers, and particularly architects, expect and are sometimes expected by their clients to adopt the style of architecture which is currently fashionable. The reason may be no more substantial than the dictates of fashion itself. So the inclusion of fashion as one of the reasons for adopting QA is not entirely frivolous. But should it be a reason? The answer lies partly in judging how ephemeral is QA. The flavour of the month? Does the organization want to admit that it does not, or is not, working towards a recognized system? It must be said that the current view, as accepted by even the most cynical of our professional journals, is that QA has now been around for the best part of a decade, and interest, although still controversial, continues to grow. Fashion may well be embedded in the sub-conscious as a more than passing reason for interest.

Market Image

Time will tell whether or not displaying a QA logo will help an organization's credibility and thus its sought-after market share of the sector in which it is operating. Many firms, whatever their motivation for investing in certification by an accredited body, lose no time in announcing their success in the professional media and in their promotional material. Thus they clearly see publicity alone as a benefit of QA. However, it is doubtful whether the designer who tends to market himself on more subjective criteria than, say, the manufacturing industry, would see much benefit in QA if the sole reason for its adoption were its perceived marketing edge.

Client Demand for QA

Here we start to explore the more important reasons for investigating the benefits of QA. Whatever its intrinsic worth to the designer, failure to pre-qualify for a commission because the client has demanded QA is a serious reason for interest. It is not known how many clients so far have made this stipulation, apart from the nuclear industry with its strict BS 5882 requirements which are wider than the demands of BS 5750. It is certain, however, that the existence (or not) of a quality system has been, and is increasingly, one of the questions a potential client will ask of design consultants before employing them.

It is certainly true that many manufacturers are now not allowed to tender unless they have the appropriate BS 5750 accredited approval. Designers therefore must anticipate that clients will increasingly be demanding that their designers practise under a nationally recognized quality system. One reason is that the clients themselves have applied QA to their own organizations and have experienced its benefits.

QA as a Means of Producing a Consistent and Disciplined Approach to Design Management

This is suggested as the principal reason for investigating QA. The other reasons, fashion, marketing or client insistence, might be admirable reasons in themselves, but they are all in a sense cosmetic. They indicate nothing of the strength of the firm's belief in the worth of good management as a means of achieving a better product. So the only lasting and uncynical reason for truly successful QA must be the inward desire that it was chosen to help the firm improve its own management. This is perhaps one reason why clients are demanding QA. They believe that their designers, however gifted, have failed to perform in some aspect of the service because management skills were seen to be inadequate.

It is therefore the philosophy of this book, that the most compelling reason for adopting QA is the wish to improve managing the processes to give the client 'assurance' that he will receive the service he was promised (the 'quality').

Is QA Appropriate for My Organization?

Here we come perhaps to the first question many readers will ask and perhaps the reason why they have persevered so far in their reading. Let us start by painting several variants of the same problem, and rhetorically ask where the reader stands. Let us imagine an organization which, however small, feels unhappy with its working methods; there is a wish for more consistency and a better structured approach to its practice. Or an organization which has been in existence for several years, and has expanded rapidly. Or an organization which may operate from several offices with increasingly difficult communication between them; there is occasional concern about its structure, are these offices becoming separate practices and if not where is the bond which holds them together? Or an organization may have made several rapid acquisitions or mergers, and may also be wondering what bonds it into one?

The feature which is probably common to all of these situations is that management has not kept pace with progress and/or expansion. The organization is still viable, still has a reputation for sound design, is still attracting clients, and is still growing. There may be any or all of these indications of success, yet the firm knows that it is practising increasingly at risk, is aware that it has a problem and is wondering what remedies are available. Readers who are asking these questions in relation to their own offices are well on the way to recognizing the benefits that the principles of QA might bring. To make absolutely sure, here are a few trial questions:

- Does the firm have a clear management structure; is it understood by all who need to understand it?
- Is the management structure capable of supporting the firm's objectives?
- Does everyone understand the firm's policy and procedures for providing the client with the quality he was promised?
- Where the firm has separate offices, is it clear whether they operate under their own procedures, centrally applied procedures or a mix? How is this relationship controlled?
- Does the firm have procedures for the critical aspects of its operation? Are they understood and followed where appropriate? Are they reviewed and kept up to date? Do the right personnel use them?

- Are the terms of engagement with the firm's clients always under control? Negotiated and signed at the appropriate level of authority? Monitored as the brief changes? Filed securely? Communicated as necessary to the team?
- Is the design process controlled throughout its stages? Reviewed? Change properly communicated and actioned? The flow of drawings always controlled?
- When the firm sublets its services, does it ensure that the process is properly controlled?
- Does the firm have a training policy?

These questions, and more, are addressed by BS 5750. If the reader feels secure after answering them, his organization probably already has a quality system which would comply with the major areas of the BS.

Developing a Strategy for Application

It is assumed that the firm will wish to proceed with developing a quality system. It is right that each firm should develop a system which respects its philosophy, its size, its workload, its mix of people and its existing quality documentation.

No book can, or should, provide a package which can be applied in its entirety. Each firm must construct its own system. Nevertheless there are principles and steps to be taken into account which, if followed, will increase the probability that a successful system will result. (See Chapters 6–19.)

To Summarize

The reasons why QA may be adopted:

- Consistency in quality of service.
- Client demand for evidence of quality systems.
- As a marketing tool.
- Fashion.

Argument that consistency in quality of service should be the principal reason for adoption.

Tests which could be applied as a means of judging whether QA might be appropriate to the firm.

6 Developing the Quality System

An organization whose principal activities are design related may not always realize that it already has a working quality system, a management structure which ensures that the service to the client is properly processed and in the right sequence. To be confronted with a set of externally imposed rules, many of which on a first reading may seem to have little relevance to design, can be a difficult introduction to QA. Some groundwork is normally required if QA is to succeed.

While published advice can be helpful, no two organizations are the same and each must develop its own route beyond a certain point. However, we indulge here in one possible scenario which it is hoped will help the newcomer to see his way towards a satisfactory introduction. It has been said many times, and is repeated here without apology, that QA will only succeed totally if it has the total and enthusiastic support of top management and in particular, where he exists, the senior principal. It is possible that the concept of QA will have been introduced by someone at a lower level in the firm who has researched and can see the benefits of BS 5750. He may well be the potential 'QA manager'.

The aim in this imagined scenario is to bring together the senior principal and QA manager in harmonious partnership; the senior principal who has the authority to drive QA towards its implementation, but quite properly may not wish to involve himself in the detail, and the QA manager who understands the detail, but may not have the authority to carry the exercise forward. So the first step is always to establish this bridgehead, wherever it is, between top management and a focus for implementation.

From the first day of commitment to QA, the process must be managed. Targets should be defined and resources allocated to meet them. There is nothing different here from running a design project. In practice and in all probability, this will require the QA manager to research what has to be done. The steps which follow should be reported in the appropriate detail to the senior principal to assure his continuing support. Programme and budgets are dealt with at the end of this chapter.

The Firm's existing Quality System and the Requirements of BS 5750: Part 1

It is important to gain an understanding of the BS and the sometimes apparently obscure intention which lies behind the words. Thorough study at this stage will avoid the need for abortive work later (similar to sound briefing and design control). It is recommended that the firm's existing procedural documentation be listed and reviewed: management structure and procedures; their relevance, how up to date they are, the extent of their usage. This will provide a basis for reviewing against the BS to appraise the extent to which the firm's existing quality documentation complies with it. It is useful to note the corresponding documentation against every clause of the BS. There will almost certainly be gaps in the documentation. When this process has been completed, there should be some indication of the task ahead. It is doubtful whether the whole task will be revealed until detailed consideration of the necessary documentation (see Chapters 7 and 8) is under way. For example, further review will be necessary to check that the procedures are auditable.

Level of Application Appropriate to the Firm

The four options to consider under BS 5750 (see Fig. 6.1) are:

Without audit

Where the firm sees BS 5750 only as a shopping list from which to select items for improving or maintaining its quality system but does not wish to commit itself to auditing. Some improvements can be made but without auditing the firm will have no assurance that the system is a complete system nor that it is being implemented. Nor will there be much feeling for how the system compares with the systems of competitors.

With internal audit but without certification

The requirement of the Standard (apart from encouragement to adopt sound quality management) is that the firm should carry out internal auditing to provide assurance that the system complies with the relevant requirements of the Standard and is being implemented. This option will enable a firm to demonstrate to a client that its quality system complies with the Standard. Implementation procedure and auditing records together with the procedures themselves will provide evidence of this.

With internal audit and with certification

BS 5750 is silent on the subject of certification. However, the

Fig. 6.1 Four options for level of application

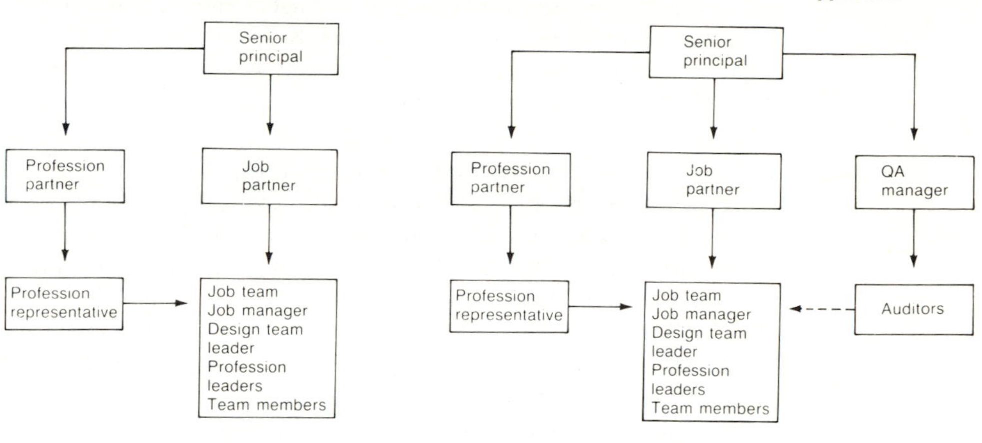

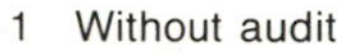

2 With internal audit but without certification

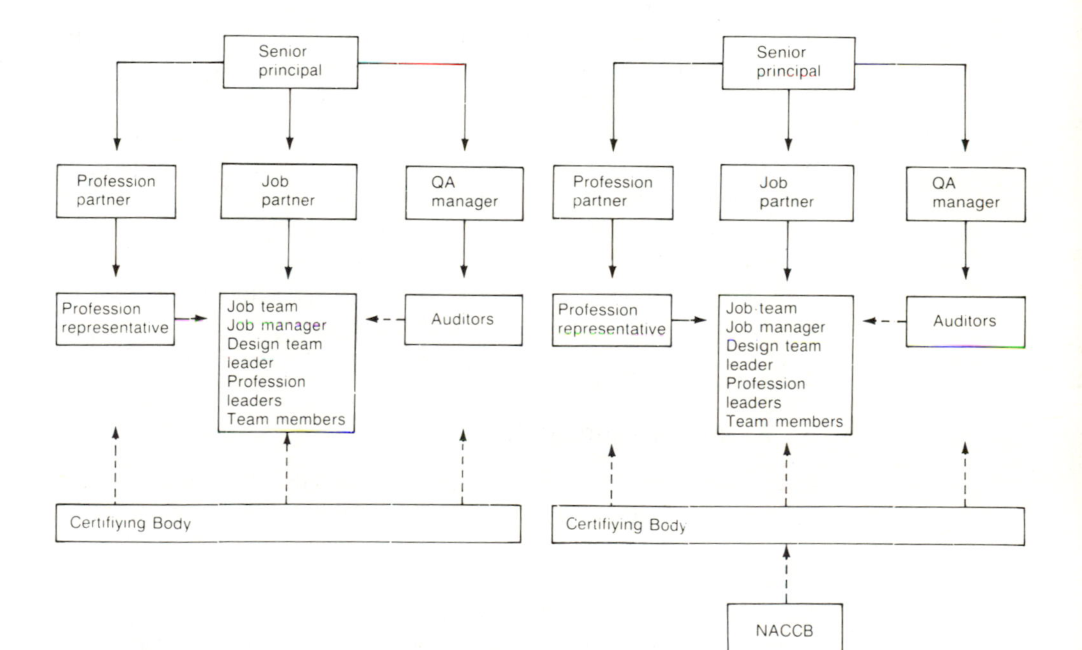

3 With internal audit and with certification

4 With internal audit and accredited certification

Government intention in the White Paper (reviewed in Chapter 4) encouraged the growth of independent certification bodies, who would bring greater consistency into the process, and by implication avoid the need for the client to investigate the quality system of every firm he appoints. This approach gives the firm a quality system which is certified as conforming to the British Standard.

With internal audit and accredited certification

This is as in the option above but where the certification body has been accredited by NACCB for the area of activities which are the subject of the certification. This has all the advantages of the above option with the added strength that the independent certification body has satisfied the NACCB that it has adequate experience of carrying out assessments of the activities which are covered by the certificate. The Ministry of Defence have waived the right to audit firms who have achieved certification by NACCB accredited certification bodies.

A further option, Total Quality Management, is discussed in Chapter 23.

Procedures

The quality of the job team's performance is a substantial element in the success or otherwise of a design organization and the team is the element most likely to react to the quality of the procedures which form the heart of the system documentation. Procedures have to be flexible enough to meet widely varying needs, yet come within the discipline of a total quality system. So it is worth testing at an early stage in the development of the system those elements of the system most likely to be experienced by the team. If the procedures prove comprehensive, workable and liked, the potential for developing a successful system is high. Procedures are also required for the support activities which make it possible for job teams to work efficiently.

Motivating Staff

At the start of this chapter it was seen as crucial that the senior principal and the QA manager be the essential forces in the successful introduction of a quality system. When strategy has been established, senior management must be informed and their commitment sought to the important decisions which have been taken.

Following this, and to avoid rumour, all members of the firm must also be informed. At this stage the QA manager himself, and perhaps the senior principal, will have the benefit of understanding the implications of BS 5750 and its associated guidance and interpretation standards. They will have talked to others in the quality management field, including certification bodies, and will have put in hand the tasks ahead. Meanwhile, all other staff who have been going about their normal duties, will have questions and possible reservations.

'All staff' is stressed because QA is not just for professional and technical staff, it reaches everyone, (see Chapters 7 to 19). However, the messages are not necessarily identical for everyone; as the functions undertaken by staff differ, so also must the nature of the messages. It is recommended therefore that 'bands' or categories of personnel be identified. For example, senior management may not need the detail, but they need to understand the strategic management aspects. Job teams on the other hand need the detail of how QA will work at team level; job team leaders will need to understand the management aspects of putting together team documentation. Support staff, secretaries and persons responsible for the firm's technical information, archiving etc, will each need to learn those aspects which affect them.

Internal Auditing

The idea of holding independent spot checks on the implementation of the firm's procedures may be novel to the firm. Internal audit is a requirement of BS 5750 to assure the senior principal that the system exists, is being reviewed, maintained and implemented. The expression 'internal' is commonly used to distinguish audit by the firm's own staff from an audit or assessment by an external organization such as a client or a certification body.

The auditing process is described in Chapter 18. It is introduced here as a reminder that it has to be planned for in the programme leading to certification body assessment, and because it is an important element in staff familiarization with the developing system. Thought should be given at this stage to finding potential auditors. There is (as yet) only a limited recruitable cadre of professional auditors who can be recruited from outside the firm. Auditors might be 'borrowed' from other practices, but that prospect may not appeal to everyone. It is more likely that auditors will have to be found from within existing staff and trained. This method has been found to be satisfactory in practice, with the first or 'lead' auditor having attended one of the many short courses (two to five days in length) commercially available.

The job team was mentioned earlier as a group worth targeting to test the acceptance rate of the developing system. If a trained audit resource is available, backed by sufficiently finalized system documentation, it might be worth considering some trial team auditing. Careful groundwork is advised, and the purpose should be explained to the team and their co-operation sought.

The Certification Body

By this time in the development of the system the firm should have some idea of the extent to which its documentation meets the requirements of BS 5750, a plan for completing its documentation, and which of the four options it has chosen. Assuming the certification route, the time has now arrived for considering the appointment of the certification body. The firm might not be ready formally to appoint the certification body. However, preliminary steps should be taken to explore which of the certification bodies can most happily be worked with, how they work, and some idea of the commercial parameters (cost and contract terms). Guidance on this process can be obtained from Chapter 4. As important a reason as any for drawing closer to the certification body to be appointed is to learn its attitude to the existing and proposed quality documentation. It may not be prepared to comment in detail until after its appointment, but should say whether it will be prepared (or wish) to comment on documentation before assessment visits.

It is not in fact advisable to rush into certification body assessment too early as this may produce the frustration of having to pay abortive charges for the privilege of being told that one's documentation is substantially incomplete. Choosing the right time is a nice balance of judgement. The results of internal auditing and early discussion with the certification body can help.

QA Manager Function Review

By this stage, we can see more clearly the remainder of the tasks required to complete the development of the quality system. Documentation is being brought together to form a total, co-ordinated system, procedures are written, rewritten or discarded. The appointment of the certification body is close. The staff understand, are motivated, and (one hopes) are supportive of the system. The time is right for internal auditing to test the implementation of the system.

It will be appropriate around this time (earlier perhaps, but not much later) to review the future quality manager role and the support he will need. It may have been determined by the senior principal at the outset that the person appointed to carry out initial research would continue to manage the development and future maintenance of the system. His function will include not only research and advice but also leadership. It will change again after implementation where the dynamic of introduction will become the auditing and review of a working system.

Completing the System Documentation

So we come to the end of one process and the beginning of another. Completion of the system documentation enables the system to be implemented. Application may be phased, i.e. as a part of the documentation is completed, it is applied. There is nothing wrong in gradual implementation and indeed, trial audit recommended earlier is an example. However, it must not become open-ended, and it will probably be found that for one part of the system to operate, other parts are required. Irrespective of how partial the introduction process is, one date should be fixed and announced to the whole firm (or such discrete parts as are identified for separate certification) on which the system will be totally implemented. After that date, all work will comply with the system, including work remaining in existing jobs, and the formal audit programme will commence. At that point, management review also starts, of which system review is an important element. By that date it is advised that the certification body should have been appointed so that it can arrange to assess the firm (or the discrete parts) for compliance with the BS.

Programme and Budget

Paradoxically, these important matters which must be actioned at the start of the process, have been left until almost the end of this chapter. This is because it was considered advisable for the reader to absorb the principal elements of introducing a system. He is now in a better position to prepare a programme and a budget.

A documented programme is an absolute necessity (as it is for any but the smallest client commission). Like all good programmes it will start with broad assumptions, and change as various elements come into focus. It should be frequently reviewed and major changes reported to the senior principal. The staff should also be regularly informed on progress, particularly the date targeted for 'going live'. A model bar-line programme, Fig. 6.2, reflects the elements of this chapter. The units of time are more likely to be months than weeks, but beyond that, realistic comment on the likely time span is difficult as so much depends on content and application of the existing quality system, and attitudes within the firm. The budget will probably be an important element, and some attempt to estimate costs will be necessary at the beginning of the process. Chapter 20 deals with some specific aspects.

Consultant Help

Consultancy advice is available to firms who do not have in-house resources to answer the many questions which arise. A consultant can help typically in:

- Explaining and interpreting the requirements of BS 5750, and associated British Standards.
- Reviewing the firm's existing documentation and advising what

Fig. 6.2 Development programme

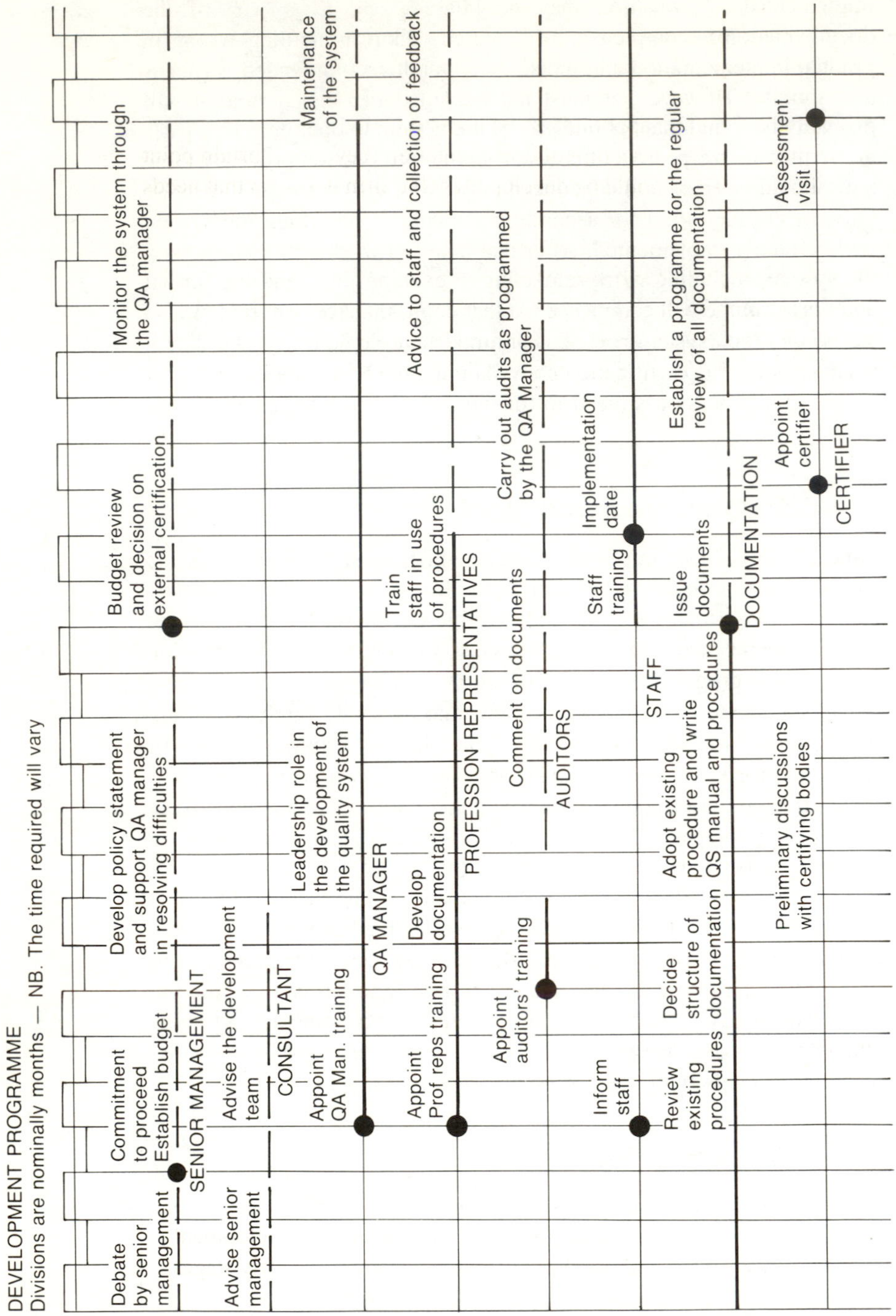

further measures are required to satisfy the Standard and the potential certification body.

- Commenting on motivation of the firm and its perceived needs of a quality system.
- Advising on the appointment of a certification body.
- Advising on the structure of documentation (beyond a certain point documentation should be developed by the firm itself, so that needs and expression are self-generated and therefore appropriate).
- Adequacy of the firm's management structure to support and manage the quality system.
- How the system can best be introduced to staff.
- A programme for implementing and operating the system.
- Auditing

The rise in popularity of consultancy advice has been stimulated by the current Government help in 'priming' or 'seeding' quality systems. Through the DTI Quality Initiative Government will fund up to two-thirds of the cost of 15 man-days consultancy. Further details are given in Chapter 20.

To Summarize

Senior Management

Establish the commitment of senior management and particularly the senior principal. This may involve key personnel attending conferences/seminars and talking to a consultant and to other organizations with experience in developing quality systems.

Obtaining any available financial support from the Department of Trade and Industry. (See Chapter 20.)

The ongoing role of senior management is to decide matters of policy and take actions to ensure the implementation of the system.

Consultant

The consultant's initial role will usually be to assess the firm's working methods and then advise how a quality system can be developed.

If senior management decide to proceed with the development of a system, he can advise the development team in the early stages of their work.

QA Manager

The QA manager should be appointed as soon as possible. He may be recruited from outside the organization but more usually this role will be taken by an experienced person from within the firm, who will commit a proportion of his time to this role. It is important that he attends a suitable training course at an early stage to ensure that he has a thorough grounding in the principles of quality systems.

Profession Representative(s)

Key personnel who have experience of the firm's working methods and adequate authority should be identified to develop the system documentation. They also should attend a suitable training course on the writing of auditable procedures.

Auditor(s)

The appointment of internal auditors can come later but it is an advantage if they are identified and trained at an early stage so they can comment on the draft system documentation, and whether it can be effectively audited.

Staff

It is important to inform all members of staff at an early stage of the decision to develop a quality system. They should be kept informed throughout and, in the period immediately prior to the implementation date, there should be specific training in the new procedures that the system will require.

Documentation

The prime responsibility for developing documentation will fall on the profession representatives, under the leadership of the QA manager. It is useful to seek the advice of a consultant in the early stages of this process but the firm should write its own procedures.

Review existing procedures and decide the structure of the quality system documentation.

The task of adapting existing procedures and writing new ones together with a quality system manual can then start in earnest.

Circulate drafts to others in the firm for comment and issue the documentation in time to allow a period of familiarisation prior to the implementation date. This is an opportunity to involve more people and gain their commitment.

Certification Body

If external certification is intended, it is prudent to make contact and negotiate with certification bodies at an early stage. This permits an informal approach to establish their interpretation of the British Standard, clarification of their fees etc, and allows a date to be established in their assessment programme, which is often committed for several months ahead. The appropriate timing of the certification body's first assessment visit should be judged on the results of internal auditing.

The firm should not be discouraged by receiving noncompliance notes and perhaps rejection on the first assessment. Certification bodies are usually able to arrange for reassessment at an early date when corrective action has been taken.

Implementation Date

When the quality system documentation has been prepared an implementation date should be established and staff advised. There must be a commitment to implementing all procedures from this date.

It is, however, not considered necessary to back-track and prepare job quality plans retrospectively covering work which has been completed.

The QA manager should set up a programme for internal auditing starting from the implementation date.

Time

The model bar-line programme for the introduction of a quality system (Fig. 6.2) identifies the various activities and how they relate.

The calendar time taken will depend on the size of the design firm and the level from which it starts i.e. the amount and appropriate uses of its existing procedures. There are two principal stages in the development of a quality system: firstly, writing the quality system manual and supporting procedures and/or adapting existing procedures; and secondly, implementation prior to assessment by an independent certification body. In the case of a multi-profession practice with seven offices, each of these two stages took approximately 15 months.

7 Writing the Procedures

Style and Structure

Just as language was essential to the development of civilization, so the written word is for QA. It is only by writing down the procedures which are to be followed and the records of actions taken, that we can demonstrate that the firm's professional activities are being properly managed. An artist is unlikely to write down the procedures by which he creates a work of fine art. In any case his client is likely to commission him purely on the basis of his reputation and past work. What is different about designing buildings? The answer is a considerable amount.

To understand this we must first look at the design process. This generally proceeds as follows:

Briefing — analysis of the design problem

↓

Postulation — of possible design solutions ←

↓

Verification — evaluation of these solutions against the brief

↓

Selection — of the preferred solution

or

Rejection and redesign → (back to Postulation)

↓

Development of the selected solution

The design of a building involves many examples of this process with new problems being identified as the design moves through its various stages.

By comparison with the artist working in his studio the building designer may require considerable organization and support systems. The skills of many individuals will often be involved and the welding of these into an efficient well-organized team is a major challenge. This is not to say that good design cannot be produced in a disorganized way but success would be hit and miss and the process inefficient. Good working methods are therefore essential if the team is consistently to give of its best and achieve its goals in a disciplined and efficient manner. QA is a means of providing assurance that this happens.

Well-organized working methods cannot provide inspiration nor are they a substitute for sound professional judgement and skill. They can, however, bring together people with the right skills, including creative talent, at the right time. They can also provide an efficient means of communication, and the many support systems which are required. Technical information, continuing professional development, feedback and equipment of various kinds are just some examples.

There is a wealth of documentation giving general guidance on working methods e.g. RIBA *Architect's Job Book* which is an excellent work of reference. QA, however, requires that, as well as general guidance, procedures should specifically state key actions to be taken, demonstrating: who does what, when, how, and, if it is not obvious, why. The how will include the method of recording actions taken. This will enable an auditor to establish, not only whether the designers have adequate working methods, but also whether they have implemented them.

As working methods vary between professions and even between different firms within the same profession, it is not often possible to write procedures which can be universally adopted. Ideally procedures should be written by the practitioners who will use them. If not, they should be written by someone who will consult those involved to ensure that the procedures are based on established good practice and that they are realistic. Obviously the procedures should give adequate control and if a BS 5750 Quality System is required, they should comply with the relevant clauses.

Procedures should not, however, be inflexible as it is impossible to anticipate all the circumstances which will arise. Flexibility is achieved by allowing changes, provided they are recorded and authorized by an appropriate person. This requires a 'change control' procedure to be established.

Procedures must cover all aspects of job running from accepting the commission to feedback and archiving. Clearly it is better to write 'standard procedures' than to expect each job team to re-invent the wheel on every job. In doing this it must be recognized that jobs differ and deviations from the standard procedures must be allowed provided that alternative procedures: are written, are authorized, meet the requirements of the Standard and enable the client's requirements to be met. Such changes should be included in a job quality plan (JQP) which also includes other job specific information. (Chapter 17 explains the concept of the JQP.)

Chapters 8 to 18 include guidance on writing procedures and include examples. It is hoped that these will help by indicating one approach but the firm should establish its own procedures which may of course take a different form.

Finally, before considering the various types of procedures it is important to recognize that it is neither possible nor desirable to write down everything which must be done in the course of doing a complex job, for example, when designing a major building. To do this would cause the wheels to grind increasingly slowly with less and less benefit and bring QA into disrepute. It is better to concentrate on those key activities which really do have potential for causing problems. Feedback will indicate what these are.

To Summarize

- QA requires procedures to be written for key activities covering: who, what, how and, if not obvious, why.
- Ideally procedures should be written by the practitioners who will use them.
- Procedures should cover the BS 5750 requirements, be based on good practice and should not be unrealistic.
- Flexibility should be allowed, subject to modified procedures being controlled and meeting the BS 5750 requirements.

8 The Quality System Manual

Introduction

Clauses 4.1 and 4.2 of BS 5750: Part 1 lay the foundations for the quality system. These clauses require that the firm:

(4.1.1) Defines and documents *its policies and objectives for, and commitment to, quality*. That it ensures that the policy is *understood, implemented and maintained at all levels* in the firm.

(4.1.2.1) Defines those who *manage, perform and verify work affecting quality*.

(4.1.2.2) Identifies verification requirements, provision of resources and assigns personnel for verification (verification should be independent inspection of the system processes which includes design review and audit).

(4.1.2.3) Appoints a management representative who will ensure that the requirements of the BS are implemented and maintained.

(4.1.3) Ensures that the quality system is reviewed at appropriate intervals.

(4.2) *Establishes and maintains a documented quality system* which complies with the requirements of the BS.

These clauses oblige the firm at its highest management levels to analyse its philosophy for ensuring that its system is an appropriate basis for it to provide the quality of service it undertakes to its clients. From this analysis should flow: formal commitment to its quality objectives; assurance that it has the management structure and a documented system which can fulfil the quality objectives; and a means of managing the system. The strategy for developing these points was set out in Chapter 6. The intention here is to advise on how appropriate documentation may be prepared which will satisfy the BS and the firm's own philosophy.

What is the Purpose of a Quality System Manual? What Should it Contain?

Neither purpose nor content is specified in BS 5750: Part 1. However, BS 4778 in Clause 11.1 defines the manual as

> A document setting out the general quality policies, procedures, and practices of an organization.

And BS 5750: Part 0: Section 0.2 in Clause 5.3.2 describes the purpose of the manual as

> To provide an adequate description of the quality management system while serving as a permanent reference in the implementation and maintenance of that system.

BS 5750: Part 1 in Clause 4.2 only requires that . . . *timely consideration needs to be given to . . . the preparation of quality plans and a quality manual* . . . This is an indication of the flexibility of the BS in its applicability to any organization or process. While there is good reason for the separate production of quality plans, the BS permits the whole of the rest of the documented system to be contained in one manual. Whether such a step is advisable is worth discussing. For very small and simple organizations it may be appropriate for just one document to contain the documented procedures required by Clauses 4.3 onwards, as well as the matters required by Clauses 4.1 and 4.2. However, there is considerable argument for the quality manual to confine its response to 4.1 and 4.2:

Intention

There is a clear separation between 4.1/4.2 (which are concerned with policy and management) and the remainder of the BS (which is concerned with detailed application).

Internal image and distribution

The firm needs a means of stating to its staff its commitment to a quality policy, and its intentions in implementing such policy. This is better produced as a short document which can (and should) be distributed widely within its organization. (Thus responding to Clause 4.1.1 of the BS . . . *shall ensure that his policy is understood, implemented and maintained at all levels* . . .) To attach to such a statement the firm's detailed quality procedures may dilute its essential message, as well as posing distribution and control problems of a wide variety of procedures, with their possible requirements for frequent change. Moreover, not all staff need to be in possession of all procedures at the same time. There is the corresponding argument that all staff should possess the manual.

External image

The firm may wish to use its manual as a means of marketing itself generally, or to give to specific clients. Its disclosure may be contractually demanded. It is unlikely that most recipients would want, or could contractually demand, the remaining system documentation. It is also unlikely that the firm would wish to make the whole of its documentation so freely available.

The remainder of this section is based on the assumption that the manual responds only to Clauses 4.1 and 4.2 of the BS. Therefore, in broad terms the purpose of the manual is to: state the firm's quality policy; explain its management structure and quality management structure; and explain how the documented quality system responds to the above.

Necessary Components of the Manual

Quality policy objectives (Clause 4.1.1 Quality policy of the BS)

The senior principal should write and sign the quality objectives of the firm whose content should include:

- The services provided by the firm to be included in the system.
- Any services provided by the firm which are to be excluded from the system.
- The firm's commitment to operate a quality system to BS 5750: Part 1.
- The firm's commitment to meeting client requirements.
- How the firm's policy is formulated and executed.
- The size of the firm and locations from which it operates.
- Balance between 'head office' control and local autonomy (where practices have more than one office).
- Legal basis of trading (e.g. limited liability, partnership).

The quality policy objectives might also cover peripheral matters which affect the nature of the firm's services to its clients. Examples are:

- Links with other organizations (e.g. single project partnerships).
- The basis on which it discharges special functions (e.g. departments or service companies which provide computing support).
- Subsidiaries or (where the firm itself is a subsidiary) the parent company.
- Policy for subletting parts of its services to others.

In the objectives the senior principal should commit himself to his part in maintaining the quality system. It should be stated here or under the management representative role (below) that the QA manager is appointed by and answers to the senior principal.

Management of the firm/office

(Clause 4.1.2.1 of the BS: Responsibility and authority; Clause 4.1.2.2: *Verification resources and personnel* is best addressed in specific procedures, and is dealt with in this way in this book.) Management is best represented by a flow diagram followed (or preceded) by an explanation of the roles, status, and authority of those shown. There should be a clear management line between senior principal and the job team who provide the 'front-line' service to the firm's clients. It might be helpful to bear in mind when preparing the diagram (particularly for larger firms where delegation is inevitable) the distinction between legal responsibilities (ownership of the firm, ultimate responsibility to client) and functioinal responsibilities (given to those who have responsibility for specific roles or tasks).

The principal support services should be shown (e.g. financial management, information provision, computing, office management) and their own line links back to the senior principal. Thus the management hierarchy might (for a firm with more than one office) be: whole firm, office and then the job team. Obviously the management structure will be simpler for the smaller firm, but the principles remain the same.

Management of the system

(Clause 4.1.2.3 Management representative of the BS.) Again a diagram is suggested followed by descriptive material.

For smaller organizations this diagram might be combined with the management of the firm/office diagram. It should show the connection between the QA manager and the senior principal. The prime function of the QA manager when the system is in operation is to oversee the audit process, so the diagram might usefully show how he devolves this function and its application at different levels of the firm's operation. If a certification body is to be appointed, its interaction with the quality system might be shown on this diagram.

Management review

(Clause 4.2 Management review of the BS.) Review of the system at appropriate intervals is required. This is best stated in a separate procedure, which might also contain the QA manager's duties. Reference to this, or these, procedures, is all that is required in the manual.

Remaining quality system documentation

(Clause 4.2 Quality system of the BS.) The firm's quality system policy and management have been described above. The manual should now address the resulting implementation of that policy — i.e. the quality system documentation. This comprises the *documented quality system procedures and instructions* referred to in Clause 4.2. How this aspect of the manual is approached depends on the firm's intended documentation structure. Figure 8.1 provides a model for the purposes of illustration only.

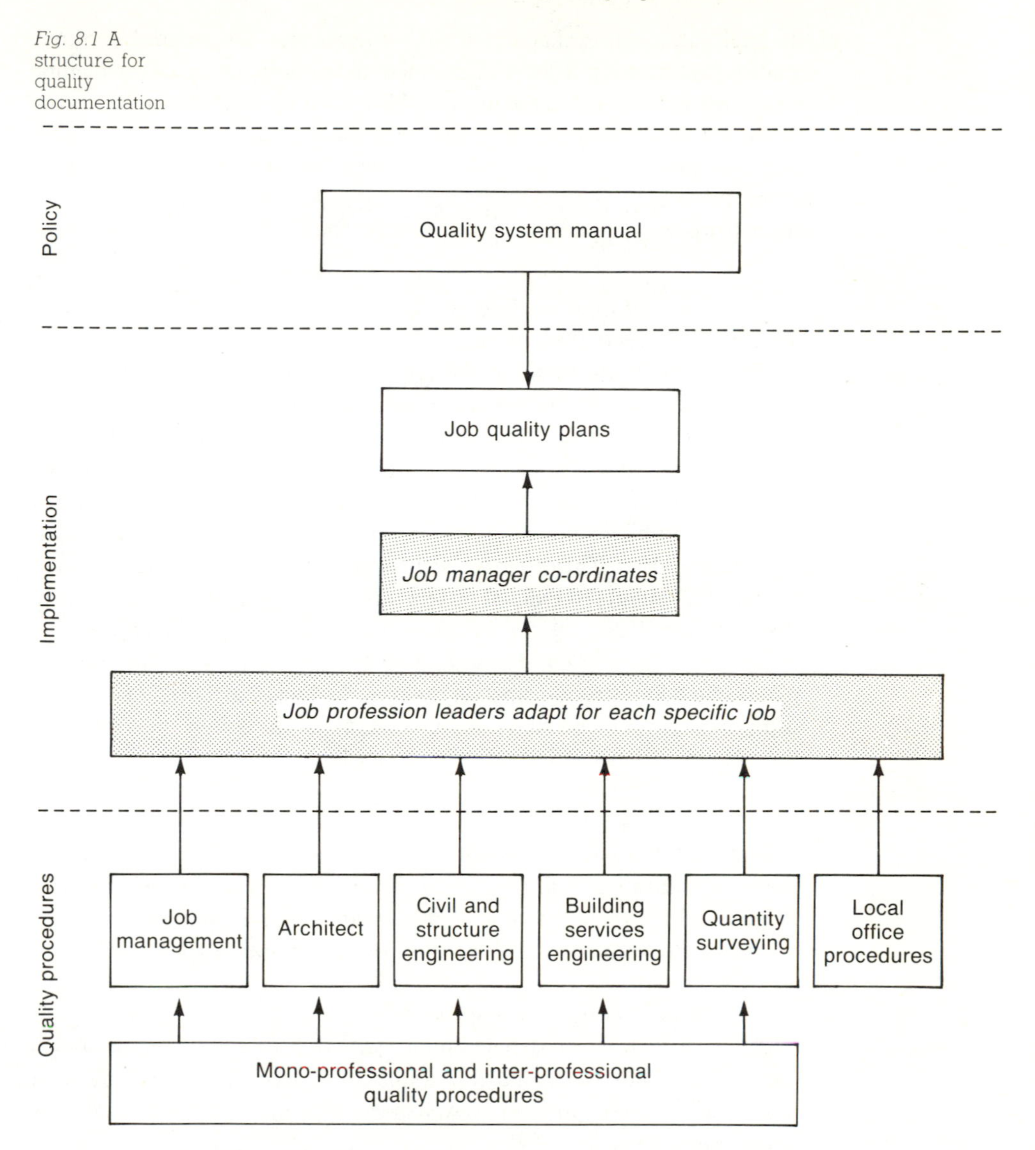

Fig. 8.1 A structure for quality documentation

To avoid the risk of turning this section of the manual into a procedure, the 'master list' of procedures should be referred to (Clause 4.5.2 of the BS requires preparation of a *master list or equivalent document control . . . to identify the current revision of documents* ...). Reference in the manual to this list is an effective way of linking manual to procedures. (The master list can then be updated as often as necessary without making the manual obsolete.)

Mention of the job quality plan (introduced in Chapter 6 and discussed in detail later in this chapter) is a most important demonstration of the quality system at job level. Commitment to the preparation of job quality plans should be stated in the manual. How their contents are to be drawn from quality documentation should also be stated.

Other Constituents of the Quality System Manual

Second and third party assessment

The client as 'second party' may have, or be encouraged to have, some right or invitation to inspect the firm's quality system (Clause 1.1 of the BS, *where a contract between two parties requires demonstration of the supplier's capability to design* ...). On the other hand, the firm may wish to include a statement to the effect that, subject to contract conditions to the contrary, the client's right to inspect the system might be limited to his examination of the manual. (This aspect is discussed in Chapter 10.) Third-party (certification body) assessment should normally avoid the likelihood of second-party inspection, but if the firm does not intend to apply for third-party approval, it might be appropriate to state its policy in regard to second-party inspection. If the firm intends to achieve third-party satisfaction, its commitment, and a brief description of the process may be stated in the manual, even though BS 5750 does not require certification body assessment.

Duty of care

Chapter 10 discusses the possibility that operation of a quality system might compromise implied duty of care in the contract with the client. It is advisable to ensure that neither the manual nor the remaining quality documentation compromise the duty of care promised the client.

Total Quality Management

Chapter 21 widens the debate on the parts of a firm's operation which may, or may not, be included in a quality system. The firm may wish to state in the quality manual, as well as the appendix (below) the activities which have been included.

Compliance with BS 5750

Finally, a most effective way of demonstrating that the firm's documentation complies with the BS is to list in an appendix to the manual all the clauses and subclauses of the BS, against which should be annotated the parts of the manual and quality documentation which address them. This should identify any clauses of the BS which are not applicable to the system. This information will be found helpful by the QA manager, the auditors (internal and certification body), and the client if he examines the documentation. An example of a typical page of such an appendix is shown in Fig. 8.2.

Quality System Application and Review

The quality system is the means by which the statement of quality policy is implemented and objectives achieved. The documented quality system should cover the management of all the activities which are necessary to give effect to the quality statement.

Clause 4.2 Quality system states:

> The supplier shall establish and maintain a documented quality system as a means of ensuring that product conforms to specified requirements. This shall include
>
> (a) the preparation of documented quality system procedures and instructions in accordance with the requirements of this International Standard;
> (b) the effective implementation of the documented quality system procedures and instructions.

Clause 4.1.3 Management review states

> The quality system adopted to satisfy the requirements of this International Standard shall be reviewed at appropriate intervals by the supplier's management to ensure its continuing suitability and effectiveness. Records of such reviews shall be maintained (see 4.16).

Procedures should be established covering the relevant requirements of the BS and any other requirements which are included in the quality system. As discussed earlier, the procedures can be incorporated in the firm's quality system manual (QSM) or produced as separate documents. In the latter case the QSM will act as a central co-ordinating document with cross-references to the supporting procedures.

Having produced the QSM and supporting procedures it is essential that they are reviewed at appropriate intervals and updated when necessary. The control of this process can be achieved by establishing a continuing programme committing the firm to reviews at one or two year intervals (see example of a typical page from a development programme, Fig. 8.3).

To Summarize

- The BS requires the firm to declare its policy and objectives for quality and to establish a documented system.
- The Quality System Manual (QSM) can contain all the quality procedures but it is usually preferable for these to be separate documents.
- The QSM should contain a policy statement — signed by the senior principal.
- The QSM should demonstrate that the firm's quality system complies with the relevant parts of BS 5750: Part 1.
- The firm is required to establish a documented system which must be maintained and subjected to recorded reviews.

Fig. 8.2 The quality system manual

Typical page

Demonstration of compliance with BS 5750: Part 1: 1987

BS 5750: Part 1: 1987 Section 4 Quality system requirements	Quality system manual (Section)	Procedure (number)
4.14 Corrective action	Sec 3.3 Sec 3.4 Sec 3.5	No 120 Sec 4 No 121 Sec 4, 5 No 122 Sec 4 No 123 Sec 3, 4, 5
4.15 Handling, storage packaging and delivery	Not applicable	
4.16 Quality records	Sec 3.3	No 120 Sec 4, 5 No 121 Sec 4, 5 No 122 Sec 4 No 123 Sec 4, 5 No 124 Sec 5
4.17 Internal quality audits	Sec 3.4	No 133
4.18 Training	Sec 2.2 Sec 2.4 Sec 3.1	No 67
4.19 Servicing	Not applicable	
4.20 Statistical techniques	Not applicable	

Fig. 8.3 Development and review in the programme for manual and procedures

DEVELOPMENT PROGRAMME FOR NEW QUALITY PROCEDURES
AND THE REVIEW OF EXISTING QUALITY PROCEDURES
AND QUALITY SYSTEM MANUAL

Please note that titles shown are abbreviated

Key

Edit Date = Edition date
Last Review = Date the document was last reviewed
Prop Review = Date of next proposed review (based on policy of review on 2 years cycle).
Action by = Initials of person responsible.
Budget = Budget for review and updating.

Procedure Number	Title	Edit Date	Last Review	Prop. Review	Action By	Comments
Quality system manual		Dec 89	—	Dec 91	BP	
14	Withdrawn — July 1989 (obsolete)	—	—	—	—	
20A	AI/SOI	Apr 81	—	May 90	RDC	To be incorporated into the new procedure number ** and withdrawn when this is available.
33A	Contract period	Sept 90	—	May 92	RDC	
46	Withdrawn — Nov 88 (obsolete)	—	—	—	—	
46A	Ext of time etc, JCT 80	Sep 90	—	May 92	RDC	
50	Withdrawn — Nov 88 (obsolete)	—	—	—	—	

9 Conditions of Engagement

Clause 4.3 (Contract review) of BS 5750: Part 1 deals with review of the contract between the purchaser (the client) and the supplier (the firm providing the design services). It requires:

- Procedures to be maintained for review of the contract and its co-ordination with other activities.
- That the supplier's requirements are adequately defined.
- That any requirements different from those in the tender (the bid) be resolved.
- That the supplier has the resources to meet the contractual requirements.

It concludes typically by requiring that records of these activities be kept.

The Significance to the Firm of Contract Review

The contract itself is significant in establishing in contractually binding terms the service the firm undertakes to provide; it would be the prime reference document by both sides if a dispute between them over any alleged under or over provision or change to the quality arose. It is, in any event, the anchor document in the job context in its definition of services to be provided. It is not of course the only quality statement, but it is the base from which the whole service is developed. It has been demonstrated that a high proportion of claims against designers arise through faulty or carelessly drafted and maintained contracts, so control of the contract is an important part of the quality system.

The Procedure

Authority

The system documentation should include a procedure which lays down who has authority to draft proposals, to negotiate with the client, to review the proposal before execution, to sign the contract on the firm's behalf, to review the contract for change, and to authorize and agree such change with the client.

Scope

The term *review* may be slightly misleading. It is intended to embrace a process which starts with the firm's bid or tender for the commission, any subsequent negotiations, the preparation and execution (i.e. signing or sealing) to produce the binding contract between client and firm, and finishes with the last agreed change to the contract. It was stated earlier that the scope also includes review to ensure that resources are adequate to perform the service.

The purpose of the BS is not concerned with the firm's commercial performance, so there is no requirement for the quality system to cover these aspects of the bid or contract. However, the firm might wish to include them in the system. There may be some contractual obligation or promise by the firm to the client which would permit the client to inspect the quality system. If this is the case, the firm will no doubt wish to keep the parts of the system which are financially sensitive separate from the remainder of the system. This must be borne in mind when structuring the system, and recognized in any arrangements made to give the client access to the system.

We will now examine each of the matters described in the section on the 'scope', for their part in the system documentation.

Bid or tender

The bid may be initiated by either client or firm, in the form of an invitation by the client to make an offer, or a direct offer by the firm. The invitation may be unwritten (say, by advertisement). The offer itself may, unusually, be unwritten. The minimum to be expected of the system at this stage is that a file should be opened to contain both the formal and informal (e.g. records of telephone conversations) aspects of these early exchanges.

Negotiation

The bid having been made, there might then follow some negotiation either to clarify, to change the service requested or offered and, no doubt, to adjust the anticipated fee. At this stage it is likely that the more detailed contract conditions, perhaps with duties owed to third parties, will be subjected to detailed examination and discussion.

The firm will find it invaluable if it has developed its own checklist of matters which require particular care and examination, either for inclusion or exclusion from the draft contract. Such a checklist would be an invaluable core to any quality system contracts procedure. Such a checklist might cover:

- Signed or sealed contract.
- A description of the purpose of the project.
- Services to be provided.
- Obligations:
  - Omissions from/additions to institute's conditions of engagement.
  - Duration of services.
  - Relationships with other consultants.
  - Programme.
  - Construction cost expectations.
  - Obligations of client.
  - Confidentiality aspects.
  - Professional indemnity matters.
  - Third-party warranties.
  - Copyright.
  - Disputes.
  - Governing laws.
- Fees and expenses:
  - Percentage.
  - Time-scale.
  - Lump sum.
  - Stage payments.
  - Value Added Tax.
- Nature of construction contract.
- Site staff.

Exchanges between firm and client both formal and informal should be recorded in the file.

Execution of the Contract

Execution of the contract is the natural outcome of the bid/negotiation process. At its simplest, it might comprise the letter offering the service confirmed by a letter in reply accepting. It is more likely, however, that it will be a somewhat larger document comprising the services, the conditions and the fee basis. It should be noted here that the BS requires any differences between client and firm to have been resolved by this stage.

An example of a typical Memorandum of Agreement is reproduced below:

Fig. 9.1 Specimen Memorandum of Agreement (based on the 1983 edition of the Memorandum Agreement of the 1981 ACE Conditions of Engagement, 4A(i)

MEMORANDUM OF AGREEMENT

BETWEEN CLIENT AND CONSULTING ENGINEER FOR ENGINEERING SERVICES IN RELATION TO SUB-CONTRACT WORKS

MEMORANDUM OF AGREEMENT made the ....

day of .... 19..................

BETWEEN ....

....

....

.... (hereinafter called "the Client")

of the one part and ....

....

(hereinafter called "the Consulting Engineer") of the other part.

WHEREAS the Client has appointed or proposes to appoint ....

....

to be the Architect for the Project and intends to proceed with ....

....

....

and has requested the Consulting Engineer to provide professional services in connection with the items stated in the Appendix to this Memorandum (referred to in this Agreement as "the Works").

NOW IT IS HEREBY AGREED as follows:—

1. The Client agrees to engage the Consulting Engineer subject to and in accordance with the Conditions of Engagement attached hereto, and the Consulting Engineer agrees to provide professional services subject to and in accordance with the said Conditions of Engagement.

2. This Memorandum of Agreement and the said Conditions of Engagement shall together constitute the Agreement between the Client and the Consulting Engineer.

3. In the said Conditions of Engagement:

   (a) the ENGINEERING DUTIES shall be as set out in Appendix 1 thereof.

(b) the method of payment for duties under Clause 6 shall be that described in Clause 12.1*, Clause 12.2* thereof.

(c) the fee referred to in Clause 12.1.1 shall be ....% of the cost of the Works.

(d) the rate or rates referred to in Clause 12.2.2(a) shall be ....

....

(e) the multiplier referred to in Clause 12.2.2(b) shall be ....

(f) the QUANTITY SURVEYING DUTIES shall include those set out in the following Clauses* of the Conditions of Engagement and the fee associated therewith shall be that referred to in Clause 13:—

7(a)	7(b)	7(c)	7(d)	7(e)
7(f)	7(g)	7(h)	7(j)	7(k)

(g) the SUPPLEMENTARY DUTIES shall include those set out in Clause 8 of the Conditions of Engagement and the associated fees shall be as follows:

(1) the percentage referred to in Clause 14(a) shall be ....%*
(2) the percentage referred to in Clause 14(b) shall be ....%*

(h) the percentages to be applied to the cost of the site staff referred to in Clause 17.1(b) shall be ....% for specifically recruited staff and ....% for seconded staff.

(j) the payment for Disbursements referred to in Clause 18 shall be

(1) reimbursed at cost*

(2) a lump sum of ....
.... (£) payable in .... equal monthly instalments.*

(3) ....% of the fees and charges due to the Consulting Engineer for all duties, as specified in Clause 18.2.*

(k) The Engineer's Principal Bank referred to in Clause 24.3 shall be ....

....

....

4. The intervals for the payment of instalments under Clause 24.1(a) of the said Conditions of Engagement shall be monthly/quarterly* reckoned from the commencement of the Consulting Engineer's appointment, and the cumulative proportions referred to in the said Clause shall be as follows:—

On completion of Work Stage 3 ....percent of the total fee

On completion of Work Stage 4 ....percent of the total fee

On completion of Work Stage 5 ....percent of the total fee

On completion of Work Stage 6 ....percent of the total fee

On completion of Work Stage 8 ....percent of the total fee

* Delete as appropriate.

5. **The additional duties to be carried out in accordance with Sub-clause 9.2(t) of the said Conditions of Engagement shall be ....**

....

....

....

....

....

....

....

PII

6. The amount of professional indemnity insurance referred to in Clause 18.1 of the said Conditions of Engagement shall be .... pounds (£........................) for any one occurrence or series of occurrences arising out of this engagement.

   This professional indemnity insurance shall be maintained for a period of .... years from the date of this Memorandum of Agreement, unless such insurance cover ceases to be available in which event the Consulting Engineer will notify the Client immediately.

   The sum payable by the Client to the Consulting Engineer as a contribution to the additional cost of the professional indemnity insurance thus provided shall be .... pounds (£........................) and such amount shall become due to the Consulting Engineer immediately upon acceptance by or on behalf of the Client of any tender in respect of the Works or any part thereof/immediately upon submission by the Consulting Engineer to the Client of his report in relation to the Task.*

LIMITATION OF LIABILITY

7. Notwithstanding anything to the contrary contained elsewhere in this Agreement, the total liability of the Consulting Engineer under or in connection with this Agreement, whether in contract, in tort, for breach of statutory duty or otherwise, shall not exceed .... pounds (£........................). The Client shall indemnify and keep indemnified the Consulting Engineer from and against all claims, demands, proceedings, damages, costs, charges and expenses arising out of or in connection with this Agreement, the Works and/or the Project in excess thereof.

AS WITNESS the hands of the parties the day and year first above written

Duly Authorised Representative of the Client ....

.... Consulting Engineer

Witness ....

.... Witness

* Delete as appropriate.

It is important that the contract be filed, and subsequently archived, in secure conditions.

Provisional Contract

Work is often undertaken on the project before the contract is executed. A finally executed contract may not be achievable until the commission is well under way. Procedures should exist which require a careful record to be maintained recording the firm's assumptions, the draft documents as they evolve, correspondence, and records of informal exchanges with the client. One suggested control measure in the event of work being undertaken during prolonged negotiation is a letter or document to the client stating the assumed contract conditions under which the firm is working until execution of the contract.

Capability to meet Contractual Requirements

Typical evidence of review is an updated barchart showing resources, or one of the many software packages available for this purpose (e.g. Open Plan). The BS calls for the firm to document its procedures for controlling this process, and for records to be kept of the periodic reviews to be undertaken. Such reviews must also be documented.

Change to the Contract

The firm will need periodically to review the contract to satisfy itself that the developing design continues to respond to what was contractually undertaken, and to incorporate any subsequent changes in the service itself. A procedure should be written which controls change (i.e. requiring regular review and documented action, so that, at any time, firm and client are in agreement about the services being given and their implications for the developing design process).

To Summarize

BS 5750 (Clause 4.3) requires controls in the contract between purchaser (Client) and supplier (the firm).

In particular it requires: procedures for review ('review' includes development of the contract) and adequate records.

The reason for importance and significance of contract review are discussed, including the contract as the prime reference document in defining quality and where disputes arise.

Constituents of a procedure in the formulation and review of contracts are given, including

- Authority to prepare and commit the firm.
- The scope of the contract as interpreted by the BS.
- The process of reaching agreement with the client including a check list of matters for consideration.
- Execution of the contract.
- Provisional contracts (devices for defining services until the contract is executed).
- Means of controlling capability to meet contractual reassessments.
- Means of controlling changes in the contract.

10 Purchasing and Subletting

Clause 4.6 (purchasing) of BS 5750 covers this subject. While its intentions in requiring formal procedures for subletting are quite clear, it is less explicit in the other matters which should be covered. The definition of 'purchasing' used as the basis for the advice given in BS 5750: Part 4 has been adopted here (guide to the use of BS 5750: Part 1, subclause 4.6.1):

> In the procurement of material or external services, whether for inclusion in the product or service or in support of activities which form part of the supplier's quality system, e.g. calibration services or specialist inspection or test activities, the supplier should ensure that such material or service is of the quality specified and that conformity can be assured.

In other words, the clause is intended to cover purchased materials or services which have a direct or indirect influence on the quality of the services (e.g. design commissions) provided by the firm to its clients. For the purpose of discussion, 'purchasing' has been taken to mean all such materials or services except sublet work, which needs a different approach. The two are separately addressed below.

Purchasing

Subclause 4.6.1 refers to *purchased product*. It is not defined anywhere in this or the other quality British Standards. (A footnote to Clause 3 of the BS states *For the purposes of this International Standard, the term 'product' is also used to denote 'service', as appropriate*.) Purchased product is assumed here to mean the materials or services the firm purchases which are necessary to enable the firm to maintain the quality of its services to the client. To accept this definition in its widest application would be to require inclusion in the quality system of a range of commodities from the firm's premises to its coffee machines.

Clearly a judgement must be made on which of the purchased materials and services need to be included in the system. Some commodities should properly be subject to the full rigour of the system, requiring formal procedures, verification and auditing. (The purchase and control of computing software, for example.) Others might demand less onerous procedures. (The purchase of the firm's pencils, for example, would hardly merit coverage in the quality system.) It might be advisable that the firm documents its intentions regarding the rigour to be applied to the different aspects, so as to limit (for example) the corresponding rigour of audit.

Clause 4.6.3 (Purchasing data) makes certain stipulations in the control of purchasing documents, most of which need to be interpreted in the context of a design service. The broad intention is that care be taken in defining in the tender documents what is required, so that firm and supplier will know with certainty what is to be provided.

Clause 4.6.4 (Verification of purchased product) complements Clause 4.6.3, in requiring verification of what is purchased against the criteria required by Clause 4.6.3. Although there may appear to be confusion between some of the terms used in this subclause and the remainder of the BS, it should be assumed here that *contract* means 'the contract between the firm and its supplier', and 'purchaser' means 'the firm'.

Examples of the more important components of such materials and services might be:

1 Design-related software

A procedure should include requirements that:

- Specification for the product is adequate.
- Product purchased conforms to specification.
- Product and hardware are compatible.
- Manufacturer's later modifications are controlled.
- Product is installed by appropriately qualified personnel.
- Users are trained and proficient in use of the product.
- Users are aware of the consequences of user modification or corruption.
- Manual verification, where appropriate, takes place.
- There is a logging system to record and trace location and status of each product.

2 Surveying and other equipment used by designers.

There should also be procedures governing purchasing, maintenance and use of such equipment.

Subletting

The BS uses the term *Subcontractors* to mean the organizations to whom the firm sublets parts of its service (conversely, 'subcontractors' are normally assumed by design organizations to be subcontractors to construction contractors). The firm retains responsibility to the client, and it may, or may not, inform the client that it has subcontracted a part of the service.

Before we examine what the BS requires, it might be useful to explore the importance of subletting in a quality system. The BS rightly holds it to be important. If the quality of the whole service to the client is to be achieved, there must be appropriate controls over the parts to be sublet so that the quality produced is to the firm's satisfaction. If the firm sublets any of its duties, control of its subletting will be an important element in its quality system.

Clause 4.6.2 (Assessment of subcontractors) requires the firm to:

- Select its subcontractors with care.
- Be satisfied that they are able to meet the requirements.
- Maintain records of acceptable subcontractors.
- Demonstrate previous performance. In practice this means ability to show documented evidence of the selection process, backed by a data-bank of subcontractors' past experience.

The BS clause refers to quality system controls. This does not mean that the potential subcontractor need have any particular quality system (e.g., one which complies with BS 5750), or even if he has, that the firm has any particular obligation to audit or assess the subcontractor's quality system.

Clause 4.6.3 and 4.6.4 have been reviewed above, for their application to *purchased product*. Save for the earlier difficulties with definition, which apply also to subletting, the subclauses (as interpreted for design activities) are sensible in their requirements that care be taken in preparing the subcontract tender documents, and for effective verification. It should be noted that, assuming *product* to be the sublet work, verification of the product either on or off site does not absolve the subcontractor from his responsibility to produce the specified product. Firms should not rely contractually on such assurance; the subcontractor may not be aware of BS 5750, and in any event this is a highly doubtful way of achieving subcontract fulfilment.

Matters for inclusion in a procedure for subletting

- Firm's subletting policy (e.g. aspects which may not be sublet).
- Restrictions imposed by the individual conditions of engagement in regard to subletting, including any requirements to inform client when subletting.
- Specification/tender documents for work to be sublet.

Many of the assessment processes for subletting, e.g. the assessment, are similar to those used in the construction procurement process; the reader should refer to the section on construction (Chapter 14) for further detail.

Definitions

There is potential for misunderstanding the intent of Clause 4.6. It is recommended that the firm's procedures include the firm's assumptions on BS 5750 scope and terminology. This will clarify understanding with the firm's certification body, who may wish to interpret the clause differently from assumptions made in this section. Definitions assumed for this section, which are recommended for inclusion in a procedure, are:

Purchasing

Purchased Product	Materials and services purchased by the firm which have an influence on the quality of services provided by the firm Clauses 4.6, 4.6.1 and 4.6.4.
Supplier	The firm Clauses 4.6.1 and 4.6.3.
Supplier	The organization supplying materials and services to the firm Clause 4.6.4.
Purchaser	The firm Clause 4.6.4.
Contract	The contract between the firm and its supplier for the supply of materials and services to the firm Clause 4.6.4.

Subletting

Subcontractor	The sublet firm Clauses 4.6.2 and 4.6.4.
Supplier	The firm Clauses 4.6.2 and 4.6.3.
Supplier	The sublet firm Clause 4.6.4.
Purchased product	Services provided by the sublet firm to the firm Clause 4.6.4.
Contract	The terms of engagement between the firm and the sublet firm Clause 4.6.4.
Purchaser	The firm Clause 4.6.4.

Purchaser Supplied Products

Clause 4.7 of the BS states:

> The supplier shall establish and maintain procedures for verification, storage and maintenance of purchaser supplied product provided for incorporation into the supplies.

This clause is referable only to manufactured products and would normally be inapplicable to articles provided by designers. Designers should be cautious if they are asked to 'verify' products provided by the client because by so doing they might unwittingly attract some responsibility for the quality of these products.

To Summarize

BS 5750 (clauses 4.6 and 4.7) demand controls in *purchasing and purchaser supplied product*. The suggested BS intentions concerning *purchased product* are discussed and separated from parts of the firm's services to the client which the firm will sublet to other organizations (sublet work). Subletting is discussed in detail, both as to interpretation of the BS subclauses 4.6.2–4.6.4 for design organizations and the application in practice of these subclauses.

Matters for inclusion in a procedure are given. It is suggested that *purchaser supplied product* would not normally be applicable to design organizations.

11 Job Documentation

The orderly approach required by a quality system is of particular value in establishing an environment conducive to efficient working.

Clause 4.5 Document control of the BS reproduced in the Section on the information system, (Chapter 16) also applies to correspondence etc. produced by the job team. Procedures should be developed to cover this and the following are typical procedures (see Chapter 13 on design and production information for the control of this type of documentation).

Incoming mail (letters, telex or fax) related to a job will be collected, date stamped and stamped with the job stamp by a person identified in the job quality plan (usually referred to as the job secretary) and then passed to the job leader (see Fig. 11.1).

JOB No.	FILE No.	
RECEIVED ABC Ptr	DATE	
DISTRIBUTION	ACTION	INFO.

Fig. 11.1 A typical mail stamp

Original documents are not to be handed direct to the addressee, unless marked 'private', 'confidential', or 'personal'. Only the addressee should open such documents.

The job leader will mark up the job stamp on each separate item indicating:

- Who is to receive copies for information only.
- Who is to receive copies and take action.
- Any additional notes and directions.
- The file reference(s).

Actioned and referenced incoming mail will then be handed back to the job secretary who will distribute photocopies and file the original as soon as possible. It should be noted that file copies of incoming facsimiles should be photocopies of the original, since fax paper deteriorates rapidly and can be too faded to read after a very short time. Also, any drawings and/or 'forms for completion' received should be registered and then distributed to job team members for action as determined by the job leader. (See Chapter 13).

Any team member responsible for action on any item of incoming mail must advise the job leader if any delay in replying is likely. Outgoing mail, telex and facsimile transmissions will be processed through the job secretary.

When members of the job team need to draft or dictate letters or other correspondence, the following procedure is to be followed:

- Draft letters etc., are to include a comprehensive reference which contains the job number and file reference and writer's initials. Enclosures are to be attached, where possible, clearly indicating to whom copies are to be circulated.
- The draft and enclosures should be passed to the job secretary for typing and then passed back to the writer for final checking and signature. The finalized signed letter or other correspondence and enclosures should be passed back to the job secretary for photocopying and despatch. Authority to sign correspondence etc should be established.

Outgoing letters should be restricted as far as possible to one subject. If this is not possible the letter should have multiple file references and a file copy will be made by the job secretary for each subject covered.

It is important to indicate on a letter if it has been faxed or sent by courier with the inclusion of 'By facsimile and post' or 'By courier' typed at the top of the page. A record of the time and date of despatch should be retained e.g. by filing the facsimile despatch slip produced by the fax machine. Correspondence sent by facsimile should also be subsequently sent by post.

All outgoing correspondence should be copied by the job secretary for the appropriate job files and for inclusion in a day file if in use (see further notes under day file). The procedure should state the distribution of outgoing mail.

Outgoing facsimiles should be accompanied by a facsimile cover sheet which duly indicates the job and file references and the total number of sheets transmitted. Originals of facsimiled documents should be posted afterwards in the normal way. Copies of both incoming and outgoing facsimiles should be taken for filing and for inclusion in the day file if in use (see further notes under day file).

Any significant oral information, e.g. by telephone, should be recorded, copied to the other party and filed. The identity of the participants, date, subject, job and file references should be identified and recorded. A copy should also be taken for the day file if one is in use.

Where a large team is involved it can be beneficial to establish a day file. This will contain copies of all incoming and outgoing mail and be circulated. The procedure should state the circulation and make provision for those receiving it to sign and date before passing it on.

The files containing the job documentation will be held in a secure location under the control of the job secretary who should maintain a current list of files.

To Summarize

Establish how the flow of job information is to be received, distributed and filed.

12 Briefing

A good brief is a clear statement of the client's objectives, and expectations of the content and quality of the finished building together with the procedural, time/cost parameters within which it is to be achieved. It should contain concise information, both at a general level and in detail, regarding each of these aspects of the client's requirements.

The BS addresses this issue in Clause 4.4.3 Design input which states

> Design input requirements relating to the products shall be identified, documented and their selection reviewed by the supplier for adequacy.
>
> Incomplete, ambiguous or conflicting requirements shall be resolved with those responsible for drawing up these requirements.

It is seldom possible to write a complete brief at the beginning of the job. However an embryo brief should be established as part of the conditions of engagement which define the service to be provided (see Chapter 9). This embryo brief should then be methodically developed as the job proceeds, with the following objectives:

- Collect information from the client in a consistent and orderly way.
- Produce information in a consistent form about widely differing functions.
- Inform the design team in general and in detail of the client's objectives, needs and expectations, including any priorities.
- Provide a basis for checking design options and the final solution.
- Provide a procedural, cost and time framework within which the design team will work.
- Provide a record of the client's agreed objectives.
- Facilitate reference for later retrieval.
- It may also form a basis for the development of an operation and maintenance manual for the completed building if this is included in the conditions of engagement.

With these objectives in mind the firm should establish a procedure (which can be varied as necessary for individual jobs) for developing the brief. The Building Research Establishment document (BRE) 'Better briefing means better buildings' provides a useful basis as it contains comprehensive well-structured check lists.

The approach will be influenced by a number of factors, principally related to client type, either from the public or private work sectors, but will also depend on the state of readiness and depth of briefing information available.

Client types in the main can be subdivided into four categories:

1 Those clients, like the Department of Health, Property Services Agency and other government departments where detailed briefing information is usually available at the time of appointment. Crown buildings, hospitals and educational buildings usually fall into this category.
2 Those clients where the brief is established by consultation and in discussion with the client's representatives usually adopting the guidance and checklists as set down in the firm's procedure. Many corporate body clients in the office sector, in the industrial sector (both research and production) and the leisure sector, fall into this category.
3 Those clients who wish to establish a 'full' brief at the outset of the job as a basis for making strategic decisions in terms of site purchase and/or site development, or for the purposes of building selection for fit out and occupation or refurbishment. For this category the procedure should be followed in the determination of an 'advance' client's brief, but should a firm commission to develop or fit out an existing building result, then the 'advance' brief obtained should be reappraised and verified at each subsequent stage of development.
4 Those developer clients where no detailed briefing is available other than the stated objective of 'maximizing the investment'. Many office, retail or mixed use developments fall into this category. For this client type it is necessary for the firm to set down a written understanding of the client's brief for his acceptance and approval.

Within these four categories, jobs will also vary in size from large to small, and from simple to highly complex in terms of functional and servicing needs with the resultant need to 'fine tune' the briefing programme to the actual job requirements.

The firm's procedure should require the following information at an early date:

- Identification of those involved including: client's authorized representative; design team leader and other consultants.
- Setting up of a briefing file or, for large jobs, files based on a classification system.
- Programme for briefing meetings, agenda and who attends.

- Format for minutes of meetings and distribution of minutes.
- Programme for the formal review of the brief and the agreement of the client to it at key stages of the job. These are usually at the end of RIBA work stages:

  B feasibility
  C outline proposals
  D scheme design

  This enables stage E detailed design to commence on a sound basis.
- A change control procedure. This should recognize that there must be some flexibility as the brief is developed but introduce a control procedure for changes which reopen questions already committed. It should require:

  A record to be kept of the reason for the change.
  A check on its effect on other parties, costs and programme.
  A check on its effect on the conditions of engagement.
  Identification of who has authority to authorize changes. Normally this is the design team leader who should of course consult those affected.
  Notification of all those affected.
  Footnote: The undermentioned warning issued by RIBA Indemnity Research Limited is a timely reminder of the importance of seeing that the client's brief is without ambiguity and it is strongly advised that the firm obtains the client's approval to the brief as it is developed through the Stages of Work.

  Ambiguities in the Client's Brief

  'In a recent legal case, the judge said it is the duty of the consultant to clarify any ambiguities in the brief.

  Please note that failure to do so or to warn the client of such ambiguity could cause you serious penalty in future years.

  Over half of all claims are attributed to architects failing to confirm their intentions at the outset. Limit the extent of your professional duty by always clarifying the brief.'

To Summarize

- Establish an embryo brief as part of the conditions of engagement.
- Develop this methodically as the job proceeds.
- Review the brief and seek the clients' agreement to it at the end of work stages B, C and D.
- Implement a procedure to control changes which reopen questions already committed.

13 Design and Production Information

The process of developing the design and producing the production information can be considered under the following headings.

- Developing the design.
- Planning the production information.
- Drawing practice.
- Specification practice.
- Verification.
- Drawings control.
- Design change control.

The BS addresses these issues under the following clause:

> 4.4 Design Control — The supplier shall establish and maintain procedures to control and verify the design of the product to ensure that the specified requirements are met

The principal requirements which amplify this general statement are set out in Clauses 4.4.2 to 4.4.6 and are summarized below.

Clause 4.4.2 Design and development planning.

- Draw up and keep updated plans that identify who will be responsible for design and development.
- Plan design/verification activities and assign qualified personnel and adequate resources.
- Identify interfaces between different groups and document, transmit and review information.

Clause 4.4.3 Design input.

- Document the client's requirements (covered in the previous section on briefing).

Clause 4.4.4 Design output

- Document the design output which shall:
  Meet the brief.

Include acceptance criteria e.g. tolerances.
Conform to regulatory requirements.
Identify characteristics of the design crucial to safe and proper functioning (fire resisting doors for example should be identified as such).

Clause 4.4.5 Design verification

- Plan, document and assign functions for verification of the design to competent personnel.
- Establish that the design meets the design input (i.e. the brief) by measures such as:
  Design reviews.
  Tests and demonstrations.
  Alternative calculations.
  Comparison with proven design if available.

Clause 4.4.6 Design changes

- Identify, document, review and approve all changes.

Clause 4.5 Document control
Clause 4.5.1 Document approval and issue

Establish and maintain authorized procedures to ensure that: the pertinent issues of documents are available and obsolete documents are promptly removed.

Clause 4.5.2 Document changes/modifications

Changes to documents to be reviewed and approved. Where practical the changes shall be identified. Maintain a list of current documents.

Clause 4.9 Process control
Clause 4.9.1 General

Plan the production processes and ensure these are carried out under controlled conditions including:
Documented work instructions.
Monitoring and control.
Approval of processes and equipment as appropriate.
Criteria for workmanship.

The BS was written for quality systems which included 'production' as well as 'design'. It would be reasonable therefore to assume that Clause 4.9 Process control is not relevant to the design process. The production of the design can be satisfied through the operation of Clause 4.4 Design control. Such interpretation should be made clear in the quality system manual, e.g. as an appendix to the manual as demonstrated by Fig. 8.2.

Developing the Design

When more than one design profession is involved it is essential to establish which one acts as 'design team leader'. This should be established when the commission is first received (see Conditions of engagement in Chapter 9). The job quality plan should specify who will take this role. In a building design commission this role would normally be taken by the architect.

A programme should be established for and design team meetings involving all professions arranged to address the following as appropriate (on multi-profession jobs this process will normally be coordinated by the design team leader):

- Briefing.
- Design sessions.
- Appointment of specialists.
- Procurement methods.
- Budget.
- Statutory consents.
- Presentations of the design to the client and other bodies.

The key to the inspirational stage of the design process is informality and it is a mistake to think that this can be highly formalized. Design sessions in which ideas are discussed freely are the power house of this stage. They are the means of generating options. Formality is introduced when the proposed option is subjected to design reviews which are held at key stages in the design process (see later section on verification).

Planning the Production Information

Co-ordinated project information (CPI) codes published by the Building Projects Information Committee provide a good source for the development of a procedure. These are:

- Common Arrangement of Work Sections for Building Works.
- Project Specification: a Code of Procedure for Building Works.
- Production Drawings: a Code of Procedure for Building Works.
- Standard Method of Measurement of Building Works (SMM7).
- A Code of Procedure for the Measurement of Building Works (SMM7 measurement code).
- SMM7 Standard Descriptions (Library of Measurement Clauses for SMM7) (HMSO).
- Co-ordinated Project Information for Building Works: a Guide With Examples.

The procedure should address the following topics.

Preparation

It is important to have a formal design review at the end of stage E (detailed design) (as well as at the end of stages B, C and D) to check the design and establish that the detailed design is complete. Ideally, this should be before commencement of stage F (production information). This is sometimes impracticable, and when it is, those elements or parts of the job for which the detailed design is not complete should be identified. Stage F can then proceed with reasonable confidence on the parts that are complete.

The following is a checklist of activities (based on the CPI code) which should be covered:

- Prepare outline schedule of drawings. (See Fig. 13.1.)
- Determine the resources required.
- Determine the arrangement of the drawings set.
- Determine the format of the drawings set.
- Decide on the manner of drawings preparation.
- Prepare schedule of drawings and resources. (See Fig. 13.2.)
- Determine techniques to be used, e.g. copy negatives, CAD.
- Decide drawings content in relation to drawings purpose.
- Determine man hours to produce each drawing.
- Prepare drawings programme.

Fig. 13.1 Outline schedule of drawings. (Reproduced from the CPI production drawings code)

BUILDING PARTS	NOTES	DRAWINGS
Floors and galleries	2nd to 4th repeated layout Mezzanine	4 plans
Stairs and ramps	3×2 type conc. 6 type timber	2 GA's, 4 details 1 GA, 2 details
Roofs	Upside down details as DHSS job	1 plan 6 details (standard)
Frame	Co-ord with struct.eng.	8 details
External wall openings	6 window types Auto entrance doors	Window schedule 20 details 7 component dwgs
	standard linings Fire checks	Door schedule 10 details

PROJECT :

PROJECT N° :

Drawing no	Title	Scale	Sheet size	Man days	Notes

Fig. 13.2 Resource schedule showing typical drawings. (Reproduced from CPI production drawings code)

- Prepare drawings register and issue to all parties using the drawings. (See Fig. 13.3.)
- Establish how the drawings will be distributed to other members of the design team (see Drawings control, Fig. 13.4).
- Provide user guide to explain the arrangement and numbering of the drawings.

Drawing Practice

Drawing standards and techniques are covered by the following British Standards:

BS 1192 Construction drawing practice
  Part 1 Recommendations for general principles
  Part 2 Recommendations for architectural and engineering drawings
  Part 3 Recommendations for symbols and conventions
  Part 4 Recommendations for landscape drawings
BS 3429 Sizes of drawing sheets
BS 4210 Specification for 35 mm microcopying of technical drawings
BS 5536 Specification for preparation of technical drawings and diagrams for microfilming

Fig. 13.3 Typical drawings schedule reproduced from CPI Production Drawings Code

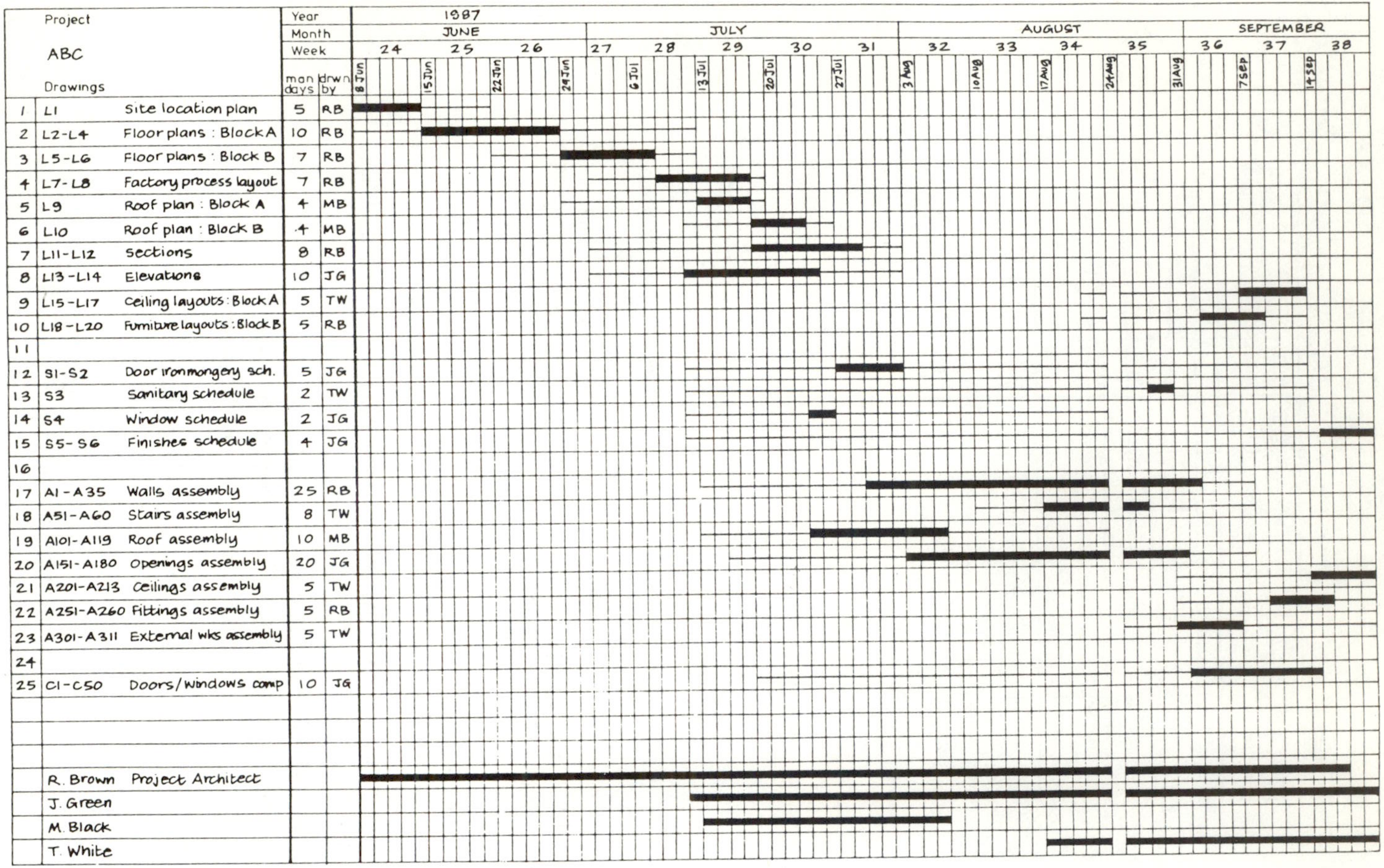

Project: ABC

Year: 1987

Month: JUNE, JULY, AUGUST, SEPTEMBER

Week: 24, 25, 26, 27, 28, 29, 30, 31, 32, 33, 34, 35, 36, 37, 38

8 Jun, 15 Jun, 22 Jun, 29 Jun, 6 Jul, 13 Jul, 20 Jul, 27 Jul, 3 Aug, 10 Aug, 17 Aug, 24 Aug, 31 Aug, 7 Sep, 14 Sep

	Drawings		man days	drwn by
1	L1	Site location plan	5	RB
2	L2–L4	Floor plans : Block A	10	RB
3	L5–L6	Floor plans : Block B	7	RB
4	L7–L8	Factory process layout	7	RB
5	L9	Roof plan : Block A	4	MB
6	L10	Roof plan : Block B	4	MB
7	L11–L12	Sections	8	RB
8	L13–L14	Elevations	10	JG
9	L15–L17	Ceiling layouts : Block A	5	TW
10	L18–L20	Furniture layouts : Block B	5	RB
11				
12	S1–S2	Door ironmongery sch.	5	JG
13	S3	Sanitary schedule	2	TW
14	S4	Window schedule	2	JG
15	S5–S6	Finishes schedule	4	JG
16				
17	A1–A35	Walls assembly	25	RB
18	A51–A60	Stairs assembly	8	TW
19	A101–A119	Roof assembly	10	MB
20	A151–A180	Openings assembly	20	JG
21	A201–A213	Ceilings assembly	5	TW
22	A251–A260	Fittings assembly	5	RB
23	A301–A311	External wks assembly	5	TW
24				
25	C1–C50	Doors/windows comp	10	JG
	R. Brown	Project Architect		
	J. Green			
	M. Black			
	T. White			

Fig. 13.4 Work/progress schedule

DETAILED SCHEDULE OF WORK / PROGRESS SCHEDULE

JOB TITLE

JOB NO | STAGE | PROFESSION

JOB PLANNER | ISSUE DATE | ISSUE NO

DRAWING NO	DESCRIPTION OF ITEM / TASK OR DRAWING TITLE	DRAWN BY	RELATED TO PROGRAMME				RELATED TO HOURS / COSTS				COMMENTS
			PLAN START	ACTUAL START	PLAN FINISH	ACTUAL FINISH	TARGET HOURS	%PROGRESS COMPLETE	TARGET HRS USED	ACTUAL HOURS	
	SPECIFICATION										
	ADMINISTRATION %										
	CONTINGENCY %										
						TOTALS	HRS	%	HRS	HRS	
							VARIANCE FROM TARGET			HRS	%

CIRCULATION: OFFICE EXECUTIVE, JOB PARTNER, JOB MANAGER, JOB PROFESSION PARTNERS, JOB PROFESSION LEADERS

Production drawings: a code of procedure for building works, mentioned previously, is complementary to BS 1192.

The above British Standards and CPI Drawings Code provide a wealth of guidance. While the designer may be tempted simply to adopt them as his procedure it is better that a drawing practice manual is established based on these sources. This allows procedures to be tailored in accordance with the designer's own working methods and to be specific e.g. on questions left as options in the CPI Drawings Code.

Note that some clients have their own requirements on drawings practice e.g. PSA who require consultants to work to the PSA 'Drawing Practice Manual'. On jobs where it is necessary to depart from your own procedure the job quality plan should state which procedure is to be used.

Specification Practice

It is essential to establish who is responsible for writing the specification. This task sometimes falls by default to the quantity surveyor even though it is not part of the service he is commissioned to provide. This should not happen as the specification is one of the principal means by which the designer develops and communicates his requirements and it is normally his responsibility.

The preparation of preliminaries is usually undertaken by the quantity surveyor, when appointed. They do however contain important information which should be checked by the designers to ensure that they meet their requirements.

The firm should establish its policy on specifying and only depart from this when circumstances on the job dictate e.g. a client such as the PSA may instruct that its own specification system is used. Departures from policy should be recorded and authorized in the job quality plan.

The National Building Specification and National Engineering Specification provide a good basis as they are periodically updated, contain guidance notes, and provide alternative sets of clauses for jobs of different size/complexity. The ICE Piling Specification and the DTI Transport Specification for highway works are examples of authoritative specifications for specific elements.

Some means of planning and controlling the process of developing the specification should be established. The control matrix (Fig. 13.5) shows how this can be done.

Verification

The firm should establish a strategy on verification recognizing that:

- It is not feasible for every decision and action to be independently checked.

Fig. 13.5 Control matrix specification system

SPECIFICATION SYSTEM - CONTROL MATRIX

JOB NUMBER   JOB TITLE

WORKSECTIONS OBTAINED FROM WP OPERATOR (Section nos)	SPECIFIER (Name)	CLAUSES MARKED UP AND RETURNED TO WP OPERATOR (Date)	CHECK BY QS ON COMPATIBILITY WITH B OF Q (Date)	WORD PROCESSING COMPLETE (Date)	CHECK BY SPECIFIER (Date)	CORRECTIONS WORD PROCESSED (Date)	FINAL CHECK BY SPECIFIER INCL ANY CROSS REFS ON DRAWS (Date & Initials)

- It is not realistic to attempt to record every step taken.
- Checks are unlikely to pick up all mistakes.
- The prime responsibility for getting the job right should rest with the job team and verification activities should not undermine this.
- A sensible balance should be struck between an over ambitious policy and *laissez faire*.
- It should pay particular attention to matters known to cause problems, feedback will indicate these.

Clauses 4.4.4 and 4.4.5 (summarized at the beginning of this chapter) cover *design output* and *design verification*.

As will be seen the BS leaves the decision on the method of *verification* open so the firm should determine this using professional judgement and having due regard to the established good practice.

Clause 4.4.2.1 of the BS requires verification to be by *qualified personnel* and Clause 4.4.5 requires personnel verifying the design to be *competent*. Auditors may interpret this as requiring procedures to define the formal qualifications which must be held by persons involved in verification. This may be appropriate for some verification activities but in most cases formalized qualifications will not be the best criteria. It is often better for procedures to state who has authority to authorize individuals to carry out specific verification tasks. This allows professional judgment to be exercised in selecting persons for particular verification activities.

Clause 4.1.2.2 requires verification to be *by personnel independent of those having direct responsibility for the work being performed.*

Design reviews

A design review is a formal examination of the design to check that it meets the client's brief, conforms to regulatory requirements and is also of an appropriate standard. Appropriate experienced personnel from outside the team should attend. The names of those attending and any follow up actions required should be recorded and signed off when actioned. Check lists can be developed from established sources e.g. 'Architects checklists' published by Henry Stewart's Publications. These are useful but should not be seen as an end in themselves. It is more important that the review is conducted by personnel with appropriate experience and commitment.

Design reviews should be held at key stages in the development of the design, usually at the end of the following work stages:

B Feasibility
C Outline proposals
D Scheme design
E Detail design.

Checking at the end of work stage F is covered below under checking drawings and specification.

Tests and demonstrations

In some situations the construction of a model or prototype will be required to check the visual or technical aspects of the design. Testing of mockups and/or work on site may also be required. As these are not normally part of the designer's basic service a recommendation should be made to the client covering cost and programme implications. Testing of work on site is of course common practice (see Chapter 14 section on construction.)

Alternative calculations

Procedures should be adopted for producing standardized calculations, where practical, to make subsequent reading and checking easier. Manual calculations should be prepared using standard calculation sheets and clearly indicate references to British Standards, Codes of Practice and other relevant documents used in the design.

The following is a typical procedure for calculations by a structural engineer:

The design calculations should be preceded by a statement of the 'basis for design', which should incorporate the following as appropriate:

- Description of the project brief and an evaluation of the requirements.
- Description of the associated external works.
- A conceptual design statement describing the form of structure, materials and environment.
- A detailed design base identifying all forms of loadings and their values, fire resistance, durability, ground conditions etc., together with material stresses to be used in the design.
- Identification of those external technical services that will form a part of the design process, whether it be specialist advice, computer analysis or design, model testing, research etc.
- Design standards that may be incorporated or adopted as part of the final design.

Any special or known requirements that may need to be incorporated to satisfy the needs of the local authorities and statutory undertakers should be identified.

All conclusions drawn within the calculations should be highlighted to facilitate accurate transfer of information to the drawings.

A calculation file should be set up for each job for all calculations relating to the job.

All calculations should be initialled and dated by the originator.

Comparison with proven design

Feedback from the previous successful use of a product, detail or complete design is one of the most reliable indicators. Care should, however, be exercised to check that the proposed use does not introduce differences which invalidate comparison.

Checking drawings and specification

When an independent check is to be carried out on a drawing or set of drawings the checker should be properly briefed as follows:

- The job no.
- The time programmed for the check.
- The type of check required e.g. dimensional, technical, co-ordination, Building Regulations, client's requirements, etc.
- The checker should be briefed on the design parameters; the codes and other guidance used in designing the element being checked.
- The drawings/specifications should be dated and referenced.
- The checker should familiarize himself with the total job sufficiently to permit the check of the particular element.
- The checker should use red, or another significant colour to make comment and write legibly.
- The checker should issue the documentation back to the team leader under cover of a note to record that the check has been carried out, and the date.
- If the work is subject to re-check, the checker should have access to the previously checked documents and verify that action has been taken on his comments.
- Such follow-up action to be checked out by a green, or other coloured mark and dated.

The check on the specification should include a check on co-ordination with the production drawings.

Drawings Control

The originals of drawings should be kept in drawing cabinets or other secure storage when not in use. Superseded drawings should be marked as superseded and stored separately until either archived or destroyed. It is good practice to maintain separate drawing registers for each source of drawings e.g. each profession, subcontractor etc. Figure 13.6 is a format which allows a record to be maintained of all revisions, who copies were issued to and the reason for the issue. A copy of the relevant page of the register can accompany the drawings. Alternatively a drawing issue sheet can be circulated with the drawings (see Fig. 13.7). The initials of the person who produced the drawing should be indicated on it. Key

Fig. 13.6 A drawings register

DRAWING REGISTER

Job Title

Job No | Classification | Sheet No | Originator

DATE

Drawing Title

DRWG NO | SCALE | SIZE | ORIGIN

Revision

Purpose of issue

I	FOR INFORMATION
C	FOR COMMENT
P	PRELIMINARY ISSUE
CC	COST CHECK
T	TENDER PURPOSES
R/A	RECEIVED FOR APPROVAL
R/R	REVISE AND RETURN
AP	APPROVAL IN PRINCIPLE
CT	CONTRACT ISSUE
V	VARIATION
RI	REISSUE
R	RECORD DRAWING
E	EXTRA COPIES
N	NEGATIVE
L	SEE LETTER ENCLOSED
S	SURVEY DRAWING

Distribution list

Number of copies

Drawings

Received from

Purpose of issue

Architects instruction No

Fig. 13.7 A typical issue sheet for pre-contract drawing

PRE CONTRACT DRAWINGS ISSUE SHEET 1

JOB NO

JOB TITLE

FILE NO

SHEET NO

DISTRIBUTION TO	NO COPIES

Copies of pre contract drawings as listed below are enclosed

PURPOSE OF ISSUE

FOR INFORMATION	☐	PRELIMINARY ISSUE	☐
FOR COMMENTS	☐	SURVEY DRAWING	☐
FOR APPROVAL	☐	TENDER PURPOSES	☐
COST CHECK	☐		☐

Any errors or omissions to be notified to originating profession

Please return drawings by ________

DRAWING NO	REVISION	DRAWING TITLE

DRAWINGS FROM

DATE

ORIGINATING PROFESSION

SIGNATURE

JOB ROLE

Fig. 13.8
Circulation sheet issued for incoming drawings

INCOMING DRAWINGS CIRCULATION SHEET

JOB NO

JOB TITLE

FILE NO | SHEET NO

DRAWINGS RECEIVED FOR COMMENT

DRAWING NO	REVISION	DRAWING TITLE

DRAWINGS RECEIVED FROM

DATE RECEIVED DRAWINGS

DATE RESPONSE REQUIRED

DATE RESPONSE ACHIEVED

CIRCULATION

PROFESSION	INITIALS	DATE REPSONSE REQUIRED BY	DATE DRAWINGS RECEIVED	DATE RESPONSE ACHIEVED	SIGNATURE

issues of drawing should be authorized (e.g. when issued for construction). Authorization should be by someone named in the job quality plan.

The procedure for dealing with drawings received for information or comment should require them to be date stamped on receipt and set out the procedure and any time limit for response. If an incoming drawing has to be circulated internally for comment an incoming drawings circulation sheet as Fig. 13.8 can be issued. The procedure should also include a suitable form of words for use when responding to organizations submitting drawing to avoid statements which would extend responsibility beyond the firm's commitment.

Design Change Control

All revisions to drawings should be properly described, dated and initialled by the person who revised the drawing. Where practicable the revision should also be identified on the drawing e.g. by using a 'cloud' drawn in pencil on the back of the negative around the revision. This cloud should be removed if the drawing is revised again so that the clouds only indicate the latest revisions. Copies of revised drawings should of course be circulated to all those affected.

In addition to the above certain changes should be subject to further control. Circumstances in which this should apply should be defined, for example: any change which would cause abortive design or building work; any change which would materially effect the programme or job cost; and any change which would re-open decisions already agreed with the client or other professions. In such cases the procedure should: require a record to be kept of the reason for the change and require a check on its effect on others, costs and programme; and specify that the change must be authorized in writing and who may do this.

To Summarize

Developing the Design

Arrange meetings to address:

- Briefing.
- Design sessions.
- Appointment of specialist.
- Procurement methods.
- Budget.
- Statutory consents.
- Presentations of the design to the client and other bodies.

Planning the Production Information

- Hold a design review at the end of stage E (detail design).
- Plan the production information.

Drawing Practice

- Produce a firm's drawing practice manual based on established codes.

Specification Practice

- Establish a firm's policy on specification practice.

Verification

Establish a verification strategy covering:

- Design reviews at key stages.
- Tests and demonstrations.
- Alternative calculations.
- Comparison with proven design.
- Checking drawings and specifications.

Drawings Control

Policy to cover:

- Secure storage.
- Identification of superseded drawings.
- Drawing registers.
- Records of distribution/reason for issue.
- Records of receipt of incoming drawings.
- Control of circulation.
- Format for drawings requiring a response.
- Authorization of issue.

Design Change Control

- All revisions to drawings to be described, dated and initialled.
- Certain changes to be subject to further control.

14 Procurement and Construction

Procurement

BS 5750: Part 1 deals with construction in Clause 4.9 Process control. However, the designer's duties will often include advising the client on procurement routes (e.g. alternative forms of contract) and on suitable building contractors. BS 5750: Part 1 does not specifically address these issues as it treats *design* and *production* as being functions of one organization, the *supplier*. The only clause approximately relevant to the designer's duties with regard to procurement is 4.6.2 Assessment of subcontractors.

The contractor will not normally be a subcontractor of the designer, so the clause does not actually cover this activity; neither is Clause 4.6.2 relevant to the designer's role in respect of the contractor's subcontractors. The clause is only applicable to those to whom the designer subcontracts (or sublets) part of the service for which he is commissioned (see Chapter 10).

It is desirable to take a pragmatic approach and work to the spirit of the BS rather than the letter when including the firm's role in construction procurement in the quality system. To achieve this the firm's procedures should cover the full extent of its duties. They should not inadvertently stray beyond them as this would be potentially very dangerous in terms of increased liability for matters outside the scope of the firm's commission. The firm should not therefore write the contractor's procedures and working methods for him. If on the other hand they are thought to be inadequate, this may be pointed out.

Methods of Procurement

The advice which the firm gives to its clients, on the options available for the procurement of design and construction services, will depend on several factors:

- The client's familiarity with the construction industry; the level of expertise within the client organization; the existence of any client's standing orders concerning procurement of consultants and contractors.

- The involvement of other advisers to the client, particularly project managers.
- The timing of the firm's appointment.
- The services commissioned from the firm by the client.

The advice required may vary according to the scope of services of each profession involved, as the following analysis of institutes' conditions of engagement shows:

Architects

The RIBA publication 'Architect's Appointment' makes two relevant references:

Under preliminary services and referring to feasibility studies in Clause 1.8, the architect will normally: 'Review with the client alternative design and construction approaches and cost implications.' This involves a limited responsibility for the architect. One infers that the client will make the choice of procurement route, perhaps with the help of other advisers in addition to the architect. However, if the architect is employed to provide additional services, under *Other Services* and referring to project management in Clause 2.38, the architect will: 'Provide management from inception to completion; prepare briefs; appoint and coordinate consultants, construction managers, agents and contractors.' This clearly involves the architect in much greater responsibility and risk.

Consulting Engineers

The various ACE Conditions of Engagement appear to limit the consulting engineers' responsibility to a similar level to the architect's normal service. The wording varies between the alternative ACE agreements but typically limits the consulting engineers' duties to: *Advise on conditions of contract relevant to the works* . . . Although the ACE Conditions of Engagement do not refer to procurement advice more specifically, the International Federation of Consulting Engineers (FIDIC) publish a standard agreement for project management. Consulting engineers providing project management services under the FIDIC Agreement are required to provide: *the management of planning, design, procurement, construction management and commissioning of a capital project*.

Quantity Surveyors

In the RICS Scale No.36 (29 July 1988) for quantity surveying services for building works, Clause 1.3 states that

> The fees cover quantity surveying services as may be required in connection with a building project irrespective of the type of contract from initial appointment to final certification of the contractor's account such as:
>
> (a) Budget estimating, cost planning and advice on tendering procedures and contract arrangements.

It is inferred that the normal QS service, therefore, is limited in a similar way to the architect's normal service and that they should join together in *reviewing* alternative design and construction approaches with the client.

Project Management

The RICS, through its publication 'The consultant quantity surveyor and project management — fees guidance' recognizes wider responsibilities for procurement advice, when quantity surveyors are employed to provide project management services. The RICS project management service includes advice on the selection of design and other consultants and consideration of contract type and contractor selection. The RICS, with the Project Manager Diploma Association of the RICS, now publish a 'Memorandum of agreement' and 'Conditions of engagement between client and project manager'. The guidance note which accompanies these publications includes a schedule of services which may be provided by project managers, including advice on the appointment of consultants and the selection of contractors.

As we have seen above, the firm may be asked by clients to provide project management services as additions to the normal services of architects, engineers and quantity surveyors. The firm may also be appointed as the client's project managers separate from or in addition to its appointments as architects, engineers or quantity surveyors.

The NEDO publication *Thinking about building 1985* identifies eight priority factors relevant to the choice of procurement route:

A Timing
B Controlled variation (flexibility to make changes)
C Complexity
D Quality level
E Price certainty
F Competition
G Responsibility
H Risk

These should be considered to establish an early understanding of the client's relative priorities and a timely recommendation on procurement routes made. Consideration can then be given to the appropriate type of contract. The 'RIBA Job book volume 2: contract administration 1988' contains guidance on procurement routes and types of contract.

Selection of Contractors/Subcontractors

The following is a suggested basis for the firm's procedures when making recommendations on suitable contractors and subcontractors. These actions should be taken and the results recorded on a job file prior to inviting tenders. This may be considered unnecessary when merely obtaining estimates.

- Consult the firm's job records system to establish if it has had previous experience of contractor or sub-contractor and if it has, consult the appropriate principal for comments on suitability for the proposed job (see later in this chapter for job records system).
- Establish the client's requirements and preferences.
- Obtain and follow up references and examples of previous similar work.
- The following questionnaire should be edited to suit job requirements and approved by the job partner before being sent to prospective contractors and subcontractors.

  Name and address.
  Date of formation (if a limited company, date of registration).
  If a limited company, state nominal and paid-up capital.
  Name of parent company (if any).
  Names of directors/partners, managers and company secretary.
  Name and address of bankers.
  Address of branch offices, factories or workshops elsewhere.
  Wholly owned or controlled subsidiary companies and nature of their business.
  Companies with which there are financial or trading links and nature of their business.
  Number of regularly employed staff in each of the following categories: supervisory (e.g. contract managers, agents, foremen); design/detailing; workshop operatives and site operatives.
  In which trades does the contractor adopt employing 'labour only' contracting.
  Which trades does the contractor sublet; name the firms normally used.
  Description of the contractor or subcontractor's organizational structure and definitions of delegated control (e.g. head office to local office, local office to site).
  Description of the contractor or subcontractor's organizational structure and proposed line responsibilities for the project.

Arrangements for staff training.
Annual turnover and maximum current value of contracts which the contractor can undertake. A copy of the latest balance sheet, or details of the capital employed.
Description of the range of contracts undertaken illustrating size, rate of completion over a comparable contract period, quality control, complexity (e.g. heavily serviced buildings).
Name two projects which are considered appropriate to this project, and which could be visited.
References which can be taken up (at least two).
Industrialized or patented building systems undertaken.
Does the firm have a quality system which complies with BS 5750 Part 1, 2 or 3?
If so has it been independently assessed. If it has, enclose a copy of the current certificate.
(It should be noted that it is not always feasible to require contractors and subcontractors to have quality systems complying with BS 5750, or to be registered as being independently assessed.)

The client's written approval should be obtained to the interview list for prospective tenderers.

Trade directories such as 'Kelly's Business Directory', and 'Kompass' cover company and financial data and are updated annually. Dunn & Bradstreet offer a service of confidential surveys of companies. Where there is any doubt about the financial status of a contractor or subcontractor, a check should be made with Companies House, which holds company accounts. On-line facilities exist to enable a check to be made on the financial record of a company.

Contractor Interviews

If interviews are to be held they should take place when there is a reasonable possibility that the project will go ahead and, having regard to the method of procurement, not so early that fundamental design information cannot be given to the contractor, and not so late that there is insufficient time for inviting tenders and the other preliminaries necessary before the contractor can be appointed. By the time interviews are arranged the form of construction contract will normally have been settled.

The client should be consulted before the interview process starts. Some clients will insist that the interview/tender list should comprise only their nominations, and if a rigid situation develops interviews might then lose some, if not all, of their point. The client's wishes must, of course, be respected; however, a part of the consultant's professional duty to his client is still to advise on the choice of the most appropriate contractor. The firm should try to avoid taking at face value the names on a client list and must make such enquiries as it considers necessary. It should also be remembered that the performance of a contractor can have a substantial influence on the design team's performance. Evaluation of this potential performance is better carried out at this stage than later when the tenders themselves are being considered. Any doubt as to the suitability of a contractor must be expressed in writing to the client.

Having agreed with the client the contractors whom the firm wish to interview, each contractor must be approached either formally or informally, to ascertain his willingness to attend. The contractor should be given information derived from or based on the checklist given below and agreed with the client before the approach is made. Contractors who indicate their willingness to participate should be asked to confirm their agreement to attend.

The following checklist may be used as a basis for the information which the contractor may need at this stage:

- Name of the job.
- Name of the client.
- Brief description of the job: its location, its function, its scope (floor area, approximate cost) and special features (e.g. brick clad, precast frame, heavily serviced).
- Special contractual requirements, e.g. normal competitive tender; negotiated contract; or management contract.

The interview agenda

It has been found to be helpful in assessing contractors to have a marking schedule and an example is given in Fig. 14.1. The basis of a typical agenda and some practical notes are given below:

- Introductions.
- Description of the scheme.
- Time scales and programme.
- Particular problems or matters of importance.
- Contractual requirements.
- Method of appointing the contractor.
- Review of the contractor's questionnaire.
- Winding up.

The team should discuss each interview immediately afterwards, if possible, to clarify points while they are still fresh in their minds.

Two questions which might help resolution are: which firms (if any) should be included on the tender list? And which firms (if any) should not be included on the tender list?

Postinterview Procedure

At the termination of all interviews, the team should decide which contractors will be recommended to the client for invitation to tender, helped, if necessary, by a marking schedule (see Fig. 14.1). They should

- Summarize in a meeting report what transpired at the interviews, including the team comments and marking (where appropriate).
- Distribute meeting report to the client, the design team and any consultants.
- Obtain the client's written approval to the proposed tender list.

Tender Action

The procedure for inviting tenders, receiving tenders, evaluating tenders and appointment of a contractor will vary according to the procurement route/type of contract and the client. It may also be affected by legislative requirements such as EEC Directives.

The following references provide guidance

Code of Procedure for Single Stage Selective Tendering 1977
RIBA Job Book Volume 1: Job Administration 1988
Architects Tenders and Contracts for Building — the AQUA Group

Job Records System

The following is a suggested list of information to be maintained in a job records system (see Chapter 16: Information system) about the contractors and principal subcontractors engaged on the firm's jobs for reference by future job teams.

- Name and address of the contractor or subcontractor.
- The type of work he has performed on one of the firm's jobs. This can be in the form of a simple coding system based on the Common Arrangement of Work Section system for specialist subcontractors with additional codes for general contractors, management contractors, etc.
- Name/job number of the job on which the contractor or subcontractor was engaged.

Some firms will wish to establish lists of 'approved' contractors and subcontractors and/or 'black lists'. This is a matter for each firm to decide but it should be remembered that any electronically stored information about other organizations is subject to the Data Protection Act 1984.

It is generally better to concentrate on capturing just the above information and leave it to future job teams to make enquiries internally

Job .... Job no. ....
Member of team ....
Date of interview ....

Fig. 14.1 A typical contractor's assessment form

SUGGESTED MARKING SCHEDULE

(N.B. The marking schedule should be devised specifically for each project; the example which follows is a copy of an actual job which emphasised considerable system construction content requiring a negotiated contract, parallel working and substantial design contribution by the contractor.)

The job team must decide its own headings and weightings for marking which will depend on the job characteristics.

Each member of the project team and any consultants present at the interview would complete a form, and all the marks be aggregated.

	Firms					
	A	B	C	D	Maximum	
1. *Design*						
Planning potential					6	
Structural potential					6	
Services acceptability					6	
Design quality					7	2
2. *Techniques*						
Heavy site work					6	
Factory work					12	
Finishings					5	
Output potential					7	3
3. *Management*						
Overall organization					3	
Responsibility of staff					6	
Quality of staff to be seconded					5	
Programming techniques					3	
Availability of other design services					4	2
4. *Organization*						
Attitude towards joint design team					4	
Attitude to contractors scheme					6	
Integration: profession/contractor					5	1
5. *Cost*						
Cost standards of similar work					8	
Attitude to proposed negotiation					7	
Experience of negotiation					5	2
TOTALS						10

General comments

---

and externally when they need further information. However, if the firm decides to maintain further and more detailed records of, for example, the capabilities of contractors, it is important to keep the records up to date. This is an onerous task and hence the proposal to limit the extent of information to be held in the job records system.

Construction

It has been stated in the previous section on Procurement that BS 5750 Part 1 does not recognize the division of responsibilities which normally exists between the 'building design' profession and the 'building contractor' because it is written for a single *supplier* responsible for *design/development and production*.

The BS simply deals with construction and installation in Clauses 4.9 and 4.10. Clause 4.9 is included below for information but it should be clearly understood that this remains for the contractor to address.

4.9 Process control

4.9.1 General

The supplier shall identify and plan the production and, where applicable, installation processes which directly affect quality and shall ensure that these processes are carried out under controlled conditions. Controlled conditions shall include the following:

(a) documented work instructions defining the manner of production and installation, where the absence of such instructions would adversely affect quality, use of suitable production and installation equipment, suitable working environment, compliance with reference standards/codes and quality plans;
(b) monitoring and control of suitable process and product characteristics during production and installation;
(c) the approval of processes and equipment, as appropriate;
(d) criteria for workmanship which shall be stipulated, to the greatest practicable extent, in written standards or by means of representative samples.

4.9.2 Special processes

These are processes, the results of which cannot be fully verified by subsequent inspection and testing of the product and where, for example, processing deficiencies may become apparent only after the product is in use. Accordingly, continuous monitoring and/or compliance with documented procedures is required to ensure that the specified requirements are met. These processes shall be qualified and shall also comply with the requirements of 4.9.1.
Records shall be maintained for qualified processes, equipment and personnel, as appropriate.

4.10 Inspection and testing

4.10.1 Receiving inspection and testing

4.10.1.1 The supplier shall ensure that incoming product is not used or processed (except in the circumstances described in 4.10.1.2) until it has been inspected or otherwise verified as conforming to specified requirements. Verification shall be in accordance with the quality plan or documented procedures.

4.10.1.2 Where incoming product is released for urgent production purposes, it shall be positively identified and recorded (see 4.16) in order to permit immediate recall and replacement in the event of nonconformance to specified requirements.

NOTE — In determining the amount and nature of receiving inspection, consideration should be given to the control exercised at source and documented evidence of quality conformance provided.

4.10.2 In-process inspection and testing

The supplier shall

(a) inspect, test and identify product as required by the quality plan or documented procedures;
(b) establish product conformance to specified requirements by use of process monitoring and control methods;
(c) hold product until the required inspection and tests have been completed or necessary reports have been received and verified except when product is released under positive recall procedures (see 4.10.1). Release under positive recall procedures shall not preclude the activities outlined in 4.10.2(a);
(d) identify nonconforming product.

4.10.3 Final inspection and testing

The quality plan or documented procedures for final inspection and testing shall require that all specified inspection and tests, including those specified either on receipt of product or in-process, have been carried out and that the data meets specified requirements.

The supplier shall carry out all final inspection and testing in accordance with the quality plan or documented procedures to complete the evidence of conformance of the finished product to the specified requirements.

No product shall be despatched until all the activities specified in the quality plan or documented procedures have been satisfactorily completed and the associated data and documentation is available and authorized.

4.10.4 Inspection and test records

The supplier shall establish and maintain records which give evidence that the product has passed inspection and/or test with defined acceptance criteria (see 4.16)

The designer's precise role in the processes of construction and installation will vary from profession to profession and from job to job so it is most important that it is properly defined in the conditions of engagement. An architect, for example, if commissioned on the basis of the RIBA's description of architect's services 'Architects Appointment' 1982* is responsible for the following during operations on site:

Contract administration
1.21 Administer the terms of the building contract during operations on site.

Inspections
1.22 Visit the site as appropriate to inspect generally the progress and quality of the work.

Financial appraisal
1.23 With other consultants where appointed, make where required periodic financial reports to the client including the effect of any variations on the construction cost.

Under the ACE conditions the engineer's role varies between different engineering disciplines.

The firm's quality system should not imply that the firm has responsibilities other than those for which it has been commissioned. For example, in discharging its role it may ask the contractor for *method statements* to check that the contractor is carrying out his duty under the building contract to control the works. If the designer thinks that these are incomplete or inadequate, he should of course say so, but should avoid doing so in a way which may change the demarcation of responsibility. In this context it is generally better to 'comment' and avoid giving 'approval' to matters which are the responsibility of the contractor. If the word 'approval' is used its meaning should be defined.

The designer may be the 'contract administrator' or merely be responsible for 'periodic inspection'. In either case he plays a significant role and this should be defined in his quality system. The following items should be covered as appropriate:

- A definition of the contractual duties of the contractor and nominated subcontractors including the following requirements:
  On-site and off-site testing including acceptance criteria, witnessing and records.
  Certification of products.
  Equipment and facilities to be provided e.g. measuring and test equipment.
  Sample panels including protection.
- The contractor's quality control and method statements.
- Notification of any apparent faults or omissions using standard proforma.
- The designer's duties during the construction stage and those of other parties.
- The functions of all members of staff responsible for contract administration and/or site inspection including the clerk of works.
- Authority for signing of architect's instructions, certificates, etc.

- Procedures for dealing with contractor/nominated subcontractor queries to avoid confusion between 'office based' and any 'site based' personnel.
- Records to be maintained, by whom, distribution and any standard proforma to be used (e.g. RIBA Job Record forms).
- Agenda and programme for site meetings.
- Documentation and equipment to be held on site.
- Arrangements for dealing with correspondence and drawings.

Inspection, Measurement and Test Equipment

Inaccuracy of inspection measurement and test equipment is not usually as great a problem for building designers as it is in the manufacturing operations for which the BS was initially developed. While it may therefore be less significant than some other aspects of the quality system it should be properly addressed.

The BS covers this in Clause 4.11 which requires such equipment to be: controlled, calibrated, and maintained. These requirements apply whether or not the equipment is owned by the firm. The procedure should require all equipment owned by the firm to be kept in a secure place and a person or persons made responsible for operating a system of recording removal and replacement of the equipment. The original copy of the calibration certificate should be kept on file and a copy of the certificate kept with the relevant item of equipment. When measuring equipment from other sources is used, it also should have a valid calibration certificate. The user should be responsible for checking that a current calibration certificate exists for any measuring equipment used, and for reporting any defects in the equipment to whoever is responsible for it. Any uncalibrated equipment should be marked as such and the firm should establish a policy to ensure that it is not used when accurate measurement is required.

To Summarize

Procurement and Construction

Activities under these headings are not provided for in BS 5750: Part 1 due to its manufacturing industry orientation. It is therefore necessary to make provision for these activities in the spirit rather than the letter of the BS in order to include them in the firm's quality system.

Procurement

Advice to the client on the options available for design and construction services should be in accordance with the conditions of appointment appropriate to the professional services within the commission.

Procedures are required to cover the full extent of the firm's duties in its role of advising the client.

Procedures should include:

- Advising the client on the options available for design and construction services available and most appropriate to the needs of the project.
- The selection of contractors with reference to the firm's job records system.
- Questionnaire to obtain the required knowledge of prospective contractors.
- Client approval of the list for interviews.
- Contractor interviews.
- Postinterview procedure.
- Client approval of final tender list.

Construction

Establish procedures to cover roles and responsibilities of the firm's representatives who would be involved during the construction period both in the office and on site.

Inspection, Measurement and Test Equipment

Procedures are required for the control, calibration and maintenance of all inspection, measurement and test equipment, as appropriate, used by the design professions employed by the firm.

* A new document, the *Standard Form of Agreement for the Appointment of an Architect* (SFA/92), has just been published by RIBA Publications Ltd, for RIBA, the Royal Incorporation of Architects in Scotland, The Royal Society of Ulster Architects and the Association of Consultant Architects. The 1982 Architect's Appointment will, however, remain available.

15 Advisory Work

Advisory work is not 'design', the principal remit of this book, but it can form a significant part of the activities of a 'designer' and indeed some specialize in it, examples include: surveys and reports; expert opinion and evidence.

BS 5750 Part 1 *covers design/development, production, installation and servicing*, and although some of its clauses are relevant, it does not cover advisory work. BS 5750: Part 8: 1991 (ISO 9004-2) Quality Management and Quality System Elements — Guidelines for Services, however, provides a basis for quality systems in respect of a wide range of services including *medical practice, catering, banking, surveying and legal services* (see Chapter 3 for details).

The following matters should be established when carrying out advisory work:

- Clients' requirements including any requirements for confidentiality i.e. the *service brief.*
- Procedure for carrying out the commission i.e. the *service.*
- Form of presentation of any reports including the giving of evidence i.e. the *delivery.*
- Review of the proposed procedures against the *service brief.*
- *Assessment* of the qualities of the service and feedback.

Other matters such as document control covered elsewhere in this book are of course relevant.

To Summarize

This covers rather specialist services required of the design professions and would include surveys and reports, and the role of expert witness.

Procedures are required for

- Establishing the brief.
- Carrying out the commission.
- Presenting the report or giving evidence.
- Reviewing the procedures against the brief.
- Assessment of service and feedback.

16 Support Systems

Information System

Information is the life blood of a design organization whose principal role is the translation of the client's requirements into a coherent design. Designers need ready access to a wide range of current information and may also occasionally need to refer to obsolete information to establish the 'state-of-the-art' at a particular time. The amount of information is growing exponentially and with it the gap between a person's knowledge and that which is available. Studies by BRE (Ref. Quality in traditional housing Vol 1: an investigation into faults and their avoidance, BRE Report 1982) have shown that many defects are caused by the failure of designers to use available information. It is therefore important that the firm has a well-organized information system.

BS 5750 covers the control of information as follows:

> 4.5 Document control
>
> 4.5.1 Document approval and issue
>
> The supplier shall establish and maintain procedures to control all documents and data that relate to the requirements of this International Standard. These documents shall be reviewed and approved for adequacy by authorized personnel prior to issue. This control shall ensure that:
>
> (a) the pertinent issues of appropriate documents are available at all locations where operations essential to the effective functioning of the quality system are performed;
>
> (b) obsolete documents are promptly removed from all points of issue or use.
>
> 4.5.2 Document changes/modifications Changes to documents shall be reviewed and approved by the same functions/organisations that performed the original review and approval unless specifically designated otherwise. The designated organisations shall have access to pertinent background information upon which to base their review and approval.

> Where practicable, the nature of the change shall be identified in the document or the appropriate attachments.
>
> A master list or equivalent document control procedure shall be established to identify the current revision of documents in order to preclude the use of non-applicable documents.
>
> Documents shall be re-issued after a practical number of changes have been made.

BS 5750: Section 0.2: 1987 Clause 17 Quality documentation and records, lists the following examples of the types of documents requiring control

- Drawings.
- Specifications.
- Blueprints.
- Inspection instructions.
- Test procedures.
- Work instructions.
- Operation sheets.
- Quality manual.
- Operational procedures.
- Quality assurance procedures.

This list is not exhaustive and it is essential that other information is available to designers and that it is properly controlled. Examples of such information will include relevant

- British, European and International Standards.
- Acts of Parliament.
- Building Regulations and supporting documents.
- Standard forms of building contract.
- Standard conditions of engagement.
- Standard specifications e.g. National Building Specification and National Engineering Specification.
- British Board of Agrément Certificates.
- Reference publications produced by BRE, CIRIA, BSRIA, TRADA etc.

The firm should establish a list of the documents which will be controlled and establish who will be responsible for them. In a small firm (up to say 50 people) a member of staff may cover this as a secondary function. Procedures should cover the recording of documents on loan. The firm can develop a complete information system itself or use one of the information services such as Barbour Index plc or Technical Index Ltd to provide part of it e.g. trade, technical and legal information which is available on microfilm.

The information system should include an *authorized* list of the names of people in the firm with authority to perform functions defined in the procedures, e.g. names of the QA manager, auditors, etc. This can form part of the firm's list of current procedures.

Job teams should always check with manufacturers or suppliers to ensure that they are working to the latest available trade literature as this changes so frequently. Copies of such information upon which design decisions are based should be placed in the job files and ultimately archived with the job information. This enables design decisions to be checked at a future time if challenged.

Consideration should also be given to the use of databases which can be searched on line. The following are examples:

ACOMPLINE. Covers urban development and planning, housing, transport, local and regional government.

ARCHITECTURE DATA BASE (RIBA). An on-line version of the '*Architectural Periodicals Index*', with additional records from the RIBA Library Catalogue.

BRIX (Building Research Establishment). Publications on building materials, structural design, environment, site management. International coverage.

COMPENDEX. (Computerized Engineering Index) Covers all aspects of engineering: civil engineering, materials, transportation, engineering management.

IBSEDEX (BSRIA). Literature on building services. Coverage includes air conditioning, electrical services, heating, plumbing, transport, security, ventilation, fire protection, noise, site and office organization, and alternative energy.

ICONDA. An international data base based in Germany which covers construction, civil engineering, architecture and town planning.

INSPEC (IEE). An electrical engineering database which contains publications in architectural CAD, architectural acoustics, civil engineering, town and country planning and air conditioning.

PLANEX. (Planning Exchange) Economic development, housing, planning, transport, public finance and management.

Records of the special interests/expertise of personnel should be maintained. The firm will also need to maintain some formal job records system which will include information on contractors and principal subcontractors.

Training

The increasing pace of change and the growth of information makes effective training particularly important.

Clause 4.18 Training states:

> The supplier shall establish and maintain procedures for identifying the training needs and provide for the training of all personnel performing activities affecting quality. Personnel performing specific assigned tasks shall be qualified on the basis of appropriate education, training and/or experience, as required. Appropriate records of training shall be maintained (see 4.16).
>
> Guidance to this clause in Part 4 states that the following steps should be taken:
>
> (a) identification of the way in which tasks and operations influence quality in total;
> (b) identification of the individual's training needs against those required for satisfactory performance of the task;
> (c) planning and carrying out appropriate specific training;
> (d) planning and organising general quality awareness programmes;
> (e) recording training and achievement in an easily retrievable form so that records can be updated and gaps in training can be readily identified.

The firm's training scheme should address the following issues:

Recruitment

Record previous training, experience and qualifications

Induction

A pack of information should be handed to new starters, covering the following: conditions of employment; quality system manual; procedures directly relevant to him; and guidance on how the firm's quality system works and how they can obtain additional information and guidance.

Year-out Students and Graduates

Appoint a mentor with responsibility for ensuring that students receive appropriate experience.

Continuing Professional Development

Appoint a senior person to assume responsibility for continuing professional development. He should establish a programme for in-house seminars, arrange attendance at external seminar and technical visits to at least cover the recommendations or requirements of the appropriate professional institution.

Information on training for the QA manager person development procedures and for the auditors is included in Chapters 6 and 20.

Personal Appraisal

Establish a regular pattern of appraisal for all personnel (usually annual) to review the record of the individual's experience and training and to establish his future needs.

Staff should be informed of the purpose and distribution of the recorded information to avoid any misunderstanding. The interviewing partner should use his own judgement on what is to be written particularly in the case of special expertise, where individuals often have misconceptions (arising from either modesty or immodesty) about the level of their knowledge.

Figures 16.1 to 16.3 show an example of a set of three staff information sheets suitable for collecting the above information. When completed they should be filed in a secure location.

While these information sheets may in themselves form an adequate personnel record it is desirable, except in the case of small firms, to transfer the information to a data base. This can provide facilities for searching on key words and can be set up with restricted access to prevent general access to personal information. *Architectural keywords* published by RIBA Publications list 3000 key words for use in architecture and the allied arts. (It should be noted that the Data Protection Act 1984 requires that people have access to information about them held on a data base.) Those assigning staff to particular functions should consult this information when making decisions.

Nonconformity and Corrective Action

Clauses 4.13.1 and 4.14 are the appropriate BS clauses for the quality system documentation to respond to, for feedback and corrective action.

Error and complaint, whilst closely related, can arise independently of each other. Error may be detected and dealt with without complaint; complaint may arise before error becomes evident, and indeed where error has not even arisen. The result of either is, or should be, corrective action and self-protection measures. As contemplated by Clause 4.14 of the BS, corrective action is for the benefit of the client. However, most design firms will see it as being important for their own protection as much as for the client's.

Fig. 16.1 Typical staff information sheet (1)

KEYTYPE
N - STARTER
A - AMENDMENT
L - LEAVER

COMPLETED BY

OFFICE

STATUS

DATE

PROFESSION

CODE

STAFF INFORMATION SHEET
OFFICE COPY

STAFF NO	SURNAME	FORENAME 1	FORENAME 2	FORENAME 3	INITIALS

SCHOOLS — DATES — EXAMINATIONS TAKEN WITH DATES AND RESULTS

FURTHER EDUCATION — DATES — EXAMINATIONS TAKEN WITH DATES AND RESULTS

PREVIOUS WORK EXPERIENCE
EMPLOYER — ADDRESS — POSITION HELD — DATES

SPECIAL SKILLS / EXPERIENCE ACQUIRED BEFORE JOINING — DATE

PROFESSIONAL TITLES — DATE — PROFESSIONAL TITLES — DATE

INTERMEDIATE OR QUALIFYING PROFESSIONAL EXAMINATIONS — DATE

ADDITIONAL QUALIFICATIONS, POST QUALIFICATION COURSES ATTENDED, PART-TIME OR NON-VOCATIONAL TRAINING — DATE

AWARDS, PRIZES AND / OR COMPETITION SUCCESSES — DATE

PUBLICATIONS / ARTICLES / LECTURES WRITTEN OR GIVEN BEFORE JOINING — DATE

Fig. 16.2 Typical staff information sheet (2)

NAME

OFFICE

DATE

STAFF INFORMATION SHEET

OFFICE COPY

FORMAL EDUCATION IN BDP EMPLOYMENT

DATES

PART-TIME DAY RELEASE ☐

INDUSTRIAL TRAINING PERIOD ☐

PROFESSIONAL PRACTICE PERIOD ☐

SUBJECTS STUDIED, EXAMINATIONS TO BE TAKEN, QUALIFICATIONS SOUGHT

DATE

TRAINING SUPERVISOR

CONTINUING PROFESSIONAL DEVELOPMENT

FORMAL REQUIREMENTS OF PROFESSIONAL INSTITUTE / COMPULSORY MINIMUM ACCREDITED STUDY TIME, STUDY REQUIREMENTS

DURING THE LAST YEAR

COURSES, CONFERENCES, SEMINARS ATTENDED, WITHIN AND OUTSIDE FIRM

TECHNICAL/PROFESSIONAL PAPERS AND BOOKS WRITTEN WITHIN AND OUTSIDE FIRM

LECTURES GIVEN WITHIN AND OUTSIDE FIRM

Fig. 16.3 Typical staff information sheet (3)

NAME

OFFICE

DATE

STAFF INFORMATION SHEET

OFFICE COPY

PROPOSED FOR NEXT YEAR

COURSES, CONFERENCES AND SEMINARS

TECHNICAL / PROFESSIONAL PAPERS TO BE WRITTEN

LECTURES TO BE GIVEN

SPECIAL VOCATIONAL OR OTHER SKILLS/INTERESTS RELEVANT TO FIRM INCLUDING EXPERIENCE OF WORKING ABROAD AND LANGUAGES

MISCELLANEOUS

DETAILS OF CONTACTS WITH SCHOOLS / CAREERS ORGANISATIONS / UNIVERSITIES / COLLEGES / LOCAL AUTHORITIES / PROFESSIONAL INSTITUTES

MEMBERSHIP OF PROFESSIONAL INSTITUTE COMMITTEES ETC.

ANY OTHER INFORMATION RELATED TO TRAINING NEEDS AND NOT GIVEN ABOVE

Professional indemnity insurance (PII) should be included in the firm's quality system. It may be demanded by the client and therefore part of the contract but in any case, insurers normally require that they be kept informed of all aspects of the error/complaint chain.

Clause 4.13.1 Nonconformity review and disposition requires procedures for:

> The responsibility for review and authority for the disposition of nonconforming product shall be defined.
>
> Nonconforming product shall be reviewed in accordance with documented procedures. It may be
>
> (a) reworked to meet the specified requirements, or
> (b) accepted with or without repair by concession, or
> (c) regraded for alternative applications, or
> (d) rejected or scrapped.
>
> Where required by the contract, the proposed use or repair of product (see 4.13.1b) which does not conform to specified requirements shall be reported for concession to the purchaser or his representative. The description of nonconformity that has been accepted, and of repairs, shall be recorded to denote the actual condition (see 4.16).
>
> Repaired and reworked product shall be re-inspected in accordance with documented procedures.

Clause 4.14 Corrective action requires that:

> The supplier shall establish, document and maintain procedures for:
>
> (a) investigating the cause of nonconforming product and the corrective action needed to prevent recurrence;
> (b) analysing all processes, work operations, concessions, quality records, service reports and customer complaints to detect and eliminate potential causes of nonconforming product;
> (c) initiating preventative actions to deal with problems to a level corresponding to the risks encountered;
> (d) applying controls to ensure that corrective actions are taken and that they are effective;
> (e) implementing and recording changes in procedures resulting from corrective action.

A policy document on the firm's intentions to ensure effective implementation of corrective action should exist, its content where necessary separated into the aspects which need to be known at different levels in the firm (at team level there is the need to report error and feedback; at senior management level is the need to negotiate claims and inform insurers).

Procedures should be written which:

- Identify potential error or complaint as early as possible.
- State actions to be taken on discovery of error or potential error.
- State actions to be taken on being notified of complaint, or of becoming aware of the likelihood of complaint.
- Inform complainant of actions proposed to respond to his complaint.
- State the processes to be adopted in dealing with the matter until it has been resolved.
- Appoint persons with given authority to deal with specified aspects of the matter.
- Record notification of the matter to your PII Insurers and the subsequent action taken.
- Verify that the appropriate action has been taken in dealing with the matter and the recording of such verification.
- Report to the PII insurer at the times and in the detail he requires.
- Obtain legal advice when appropriate.
- Notify by means of the feedback system the relevant aspects of the matter in order to avoid recurrence; similarly, receive by means of the feedback system and process as above, information in order to avoid recurrence.
- Ensure that confidentiality is maintained at appropriate levels in the firm. (This aspect may need to be considered when developing the audit process; it may be required that confidential information should not be made available for normal audit but subject to audit by a specific person or persons.)
- Retain 'state-of-the-art' and other documents relative to the origin of the suspected or alleged fault for future reference.
- Circulate a periodic questionnaire to senior staff to ensure that they report matters likely to be complaints, and that current intelligence is maintained.

Focus for Feedback

Feedback is essential to the successful operation of any quality system. Its collection, evaluation, and resulting action should comprise a coherent and complete system which is effectively controlled. To be complete the procedure should identify the source, focus and the methods of distribution for all the various types of feedback.

In a small firm it may be appropriate to identify one person as the focus for all feedback. This person would be responsible for recording, evaluating and directing it to others as appropriate for action and may be the person responsible for the firm's technical information. In a larger

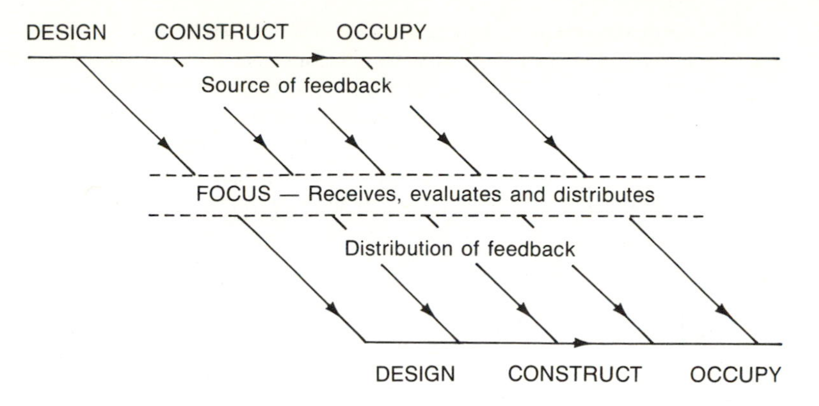

Fig. 16.4 Feedback system

organization it might be better to identify a focus appropriate for each of the different types of feedback as follows:

Subject

- Quality system manual
- Procedures
- Professional/technical information
- Services and fees
- Building contracts
- Claim/potential claims
- Office administration

Archives

The long term storage of records is important for a number of reasons: For use where information is sought about the design, for example, where further work is required to past jobs. For use on similar future jobs. For reference when disputes arise, and finally to demonstrate the effective operation of the quality system.

Clause 4.16 Quality records states

> The supplier shall establish and maintain procedures for identification, collection, indexing, filing, storage, maintenance and disposition of quality records.
>
> Quality records shall be maintained to demonstrate achievement of the required quality and the effective operation of the quality system. Pertinent sub-contractor quality records shall be an element of these data.

> All quality records shall be legible and identifiable to the product involved. Quality records shall be stored and maintained in such a way that they are readily retrievable in facilities that provide a suitable environment to minimize deterioration or damage and to prevent loss. Retention times of quality records shall be established and recorded. Where agreed contractually, quality records shall be made available for evaluation by the Purchaser or his representative for an agreed period.

The firm's archive procedures should cover the following areas:

- Who is responsible for archiving. Usually the job team leader is responsible for assembling the job documentation and passing it to the person who has responsibility for the firm's archives.
- When should this be done. It is good practice to remove from the work spaces documentation for all jobs on which work has been completed. The following criteria of judgement will determine those jobs to be moved into archives:
  - Jobs completed for which the final certificate has been issued.
  - Jobs for which a partial service only has been given and for which the fees have been paid in full.
  - Aborted jobs.
  - Competition entries when the results have been published.
- What should be retained? This usually includes copies of the following, but the final selection should be made by the job manager:
  - Memorandum of agreement and schedule of services and fees.
  - List of the firm's fee invoices.
  - Fee calculations and final fee sheets.
  - The signed forms of building contract.
  - The priced bills of quantities and specifications.
  - The firm's drawings, including: signed contract drawings; subcontract drawings, and 'as built' drawings and negatives (including those produced on CAD).
  - Signed forms of warranty.
  - Final account plus any post contract correspondence affecting it.
  - Consultants'/subcontractors' drawings.
  - Engineering calculations and bending schedules.
  - Architect's instructions.
  - Minutes of monthly progress meetings.
  - Clerk of works weekly progress reports.
  - All contractual certificates.
  - Job quality plan.
  - Job files.
  - Technical product information relied on for design decisions.
  - Site records.

and nonjob documentation such as:

— Quality system manual.
— Procedures.
— Audit reports.

- How long should the records be retained? The period should cover any period agreed with the client and the period of potential liability. This varies in the different countries e.g. Scotland and England differ on periods of liability. 20 years will cover most situations. In the case of microfilm this can be retained indefinitely as the space taken is minimal. 98 pages of A4 can be reduced to a single postcard sized microfiche.
- Prior to microfilming, consideration should be given to keeping copies of key documents in fireproof storage. Alternatively, optical discs can be used which have a large capacity and the facility for rapid retrieval of information.
- Method of indexing, to facilitate retrieval and identifying the job name/number and other key information such as building type.
- Method of storage facilites, to prevent loss or damage. Microfilm has advantages in making it relatively cheap to maintain duplicate records in two separate locations making for maximum security against loss, for example, by fire. It is usual to keep the original copy of the Memorandum of Agreement and Schedule of Services and Fees and the signed set of contract documents even when these have been microfilmed.
- Procedure for booking out, in order to control documents withdrawn from archives.
- Procedure for recording documentation passed to the client. If the fees for the work have been paid the contract documentation is legally the property of the client, irrespective of the copyright of the design.

Figures 16.5 to 16.7 illustrate an approach to archiving. Figure 16.8 is a typical microfilm index form. Figures 16.9 and 16.10 are typical archive entry control forms. Figures 16.11 to 16.14 are for obtaining and recording client instructions on the disposal of original copies of microfilmed drawings.

To Summarize

Information System

The firm is required to establish and maintain procedures for the control of all documents and data used in the operation of its quality system and which relate to the requirements of the International Standard.

Training

Procedures required to identify training needs and policy for implementation, including recruitment, induction, part-time training, continuing professional development and regular appraisal of personnel.

Nonconformity and Corrective Action

It is essential to have a controlled system of feedback which identifies source, focus and the distribution of the different types of feedback.

- Procedures are required to deal with the actions to be taken by people at all levels of the firm from job teams to senior management.
- Personnel should be identified to carry out the necessary actions.
- Verification and records should be kept of actions taken.
- Insurers should be informed of complaints or situations likely to result in a complaint.

Archives

Archive procedures are required to cover: who is responsible for archiving; when should committal of documentation to archives be carried out; what should be archived; and an outline of the archiving processes involved and the method for retrieval.

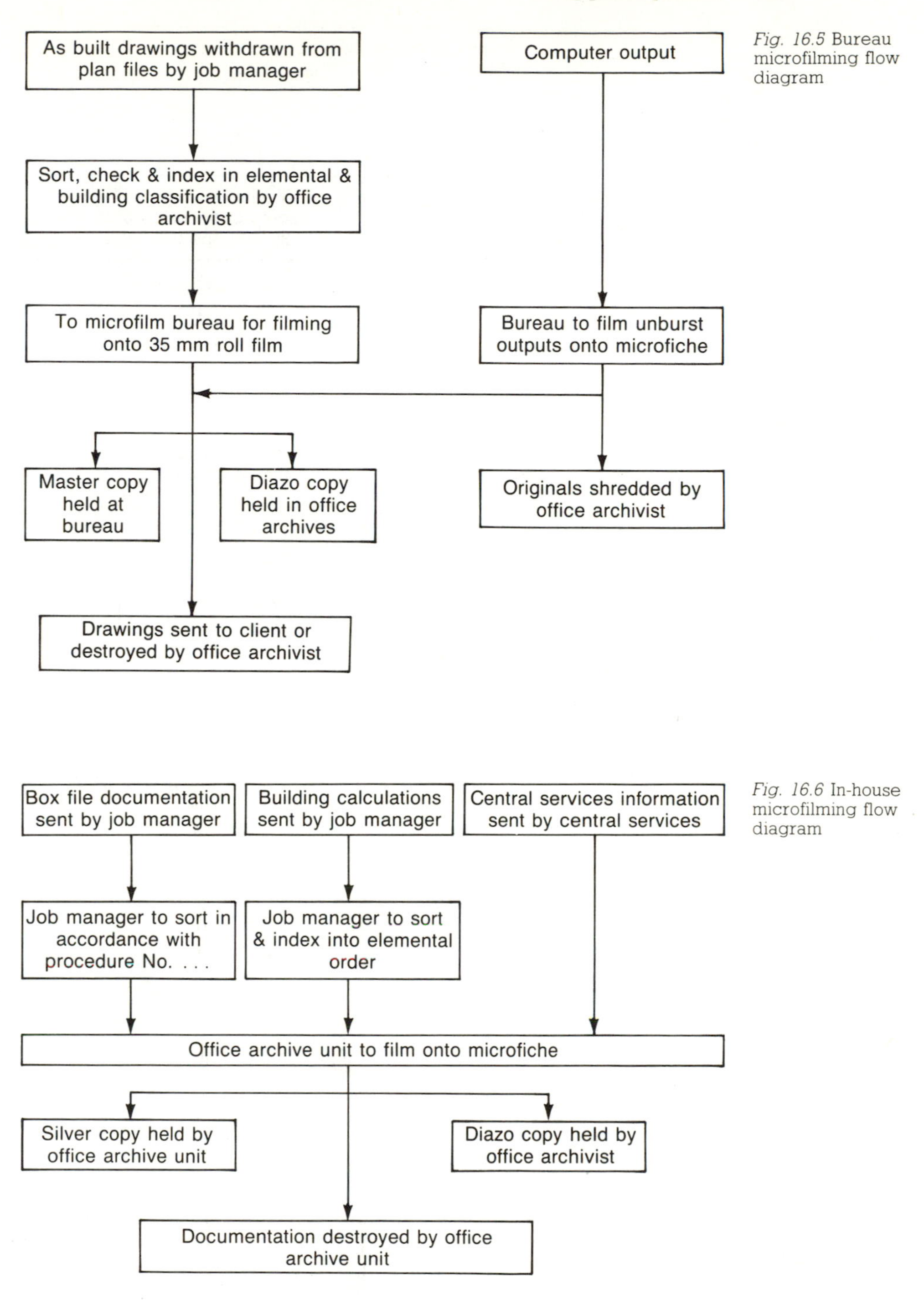

Fig. 16.5 Bureau microfilming flow diagram

Fig. 16.6 In-house microfilming flow diagram

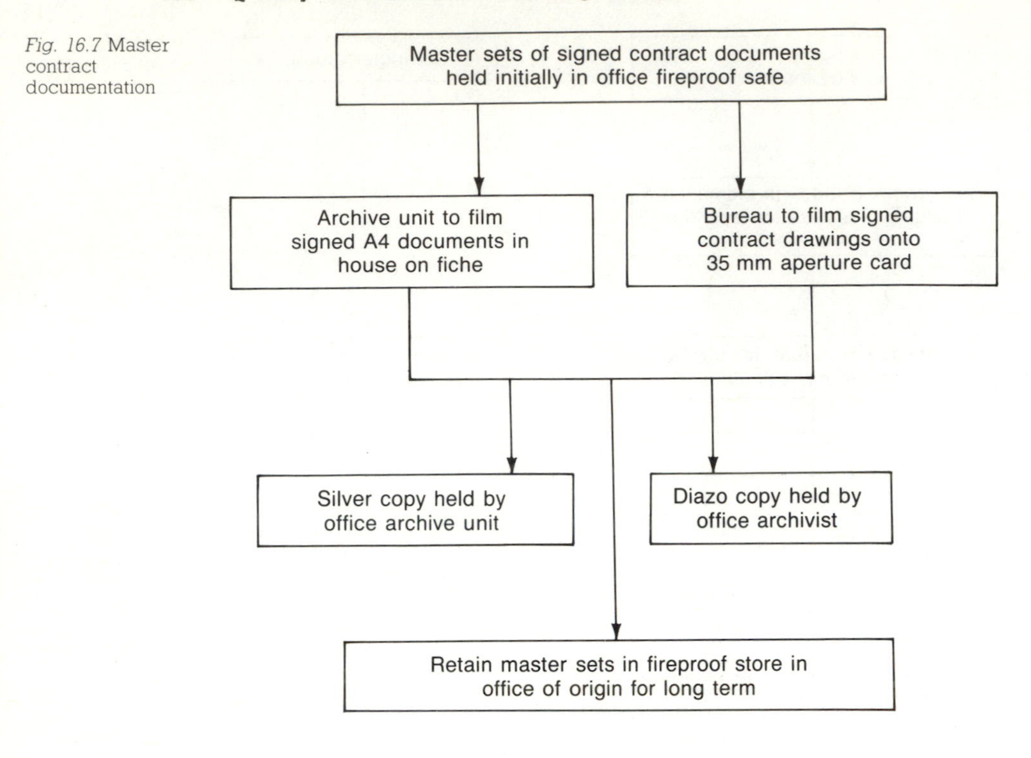

Fig. 16.7 Master contract documentation

Job No Job title

Section

Film No

Frame No	Drawing No	Description	Remarks

Fig. 16.8 Microfilm index

Fig. 16.9 Archive entry form for box files

DATE

INSTRUCTED BY

APPROVED BY JOB MANAGER

ARCHIVE ENTRY FORM

Please note that documentation will not be accepted into Archives until it has been sorted in accordance with QP93.
Once accepted by the Archivist they will be held initially in the Holding Store and thereafter Microfilmed as time allows.

			ARCHIVE USE ONLY
JOB NUMBER	NUMBER OF BOXES	CONTENTS	ARCHIVE REFERENCE

ARCHIVIST SIGNATURE

Fig. 16.10 Archive entry form for drawings

DATE

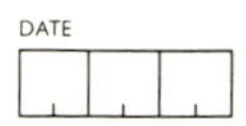

INSTRUCTED BY

DRAWINGS

ARCHIVE ENTRY FORM

APPROVED BY JOB MANAGER

Please note that drawings for microfilming will not be accepted into Archives unless they have been sorted in accordance with QP93 and are accompanied by an up-to-date drawing list/drawing issue schedule or microfilm index sheet as appropriate.

				ARCHIVE USE ONLY	
JOB NUMBER	JOB TITLE	ROLL NO/ ELEMENT NO	NO OF DRAWINGS	ARCHIVE REF NO	DATE RETURNED TO OFFICE OF ORIGIN

ARCHIVIST SIGNATURE

DATE RECEIVED

Fig. 16.11 Letter requesting client instructions on the disposal of microfilmed drawings

Dear

We have a number of drawings in our archives that relate to projects we have done for you in the past and these are listed on the sheet(s) attached. These have now been microfilmed onto 35mm roll film and the negatives are held in secure conditions.Large-scale prints can still be produced from these films when they are needed.

This makes the original drawings redundant and we are making arrangements to destroy them.Before we do so,if you would find them useful for your own purposes we should be happy to present,either these originals,at our own expense,or alternatively copies of the related microfilms at the appropriate cost.

Will you please let us know whether or not you wish to accept our offer by completing the enclosed form and return it to us before.........

Yours faithfully

Office Manager

Fig. 16.12 Typical client response form

TO:

1. I would/would not* like to take advantage of your offer dated............

2. If yes, please send:

(a) The original negatives*

(b) Copies of microfilm @ cost

* PLEASE DELETE AS APPROPRIATE

3. Please send these to the following address:-

marked for the attention of:

4. This letter is your authority to release the drawings.

Signed.................................. Date......................

Name in block letters......................

Address..................................

....

....

Fig. 16.13 Typical acknowledgement form for client response

Dear

In accordance with your request we are forwarding under cover of this letter the original drawing negatives/copies of the microfilms for the undermentioned projects:

Job No.	Job Title

We are also sending for your retention copies of the Microfilm Indexes for these projects which will assist should you have need for additional prints/negatives taking from our master set of microfilms.

Will you please sign the attached receipt as proof of the delivery and either return it by post or hand it to the delivery courier as appropriate.

Yours faithfully

Office Manager

Fig. 16.14 Acknowledgement form for drawing negatives

TO:

FROM..............................

....

....

....

I hereby acknowledge receipt of the drawing negatives/copies of microfilm relating to the undermentioned projects:

Job Number	Title	Number of Negatives

Signed.....................Position..................Date..............

17 The Job Quality Plan

The job quality plan (JQP) includes job specific information and identifies the procedures which should be implemented by the job team in order to comply with the quality system and to effectively discharge the commission from the client. Preparation of the JQP commences as soon as is practicable after receipt of the client's instruction to proceed with the commission has been received.

Clause 4.2 of the BS requires a *documented system* and the relevant guidance in BS 5750 Part 4 emphasizes that the quality plan

> . . . needs to be as complex and extensive as the project, product or service demands. It could be as simple as a document containing one paragraph or a drawing suitably dimensioned and inscribed.

Procedures should identify who should authorize the JQP and the minimum contents of the first issue. The following may be taken as a guide:

- Job number.
- Job title.
- Date of the JQP.
- Distribution of the JQP.
- Service to be provided with facility for signing this off and dating as stages are complete (with reference to file with further information including record of the contract review/authority to accept the commission).
- Description of the job (with reference to the file containing the brief).
- Reference to the file where the job programme is maintained.
- Names of the job team including the job secretary.
- Name, address, telephone/fax numbers and names of contact person(s) for organizations such as: client; building users; other consultants; local authorities; statutory authorities; etc.
- Confirmation that the standard office procedures will be used or alternative procedures defined (where permitted by the quality system).
- Statement signed and dated by a person who has authority to authorize the JQP. This is usually the partner with particular responsibility for the job.

The following additional items should also be covered on jobs where relevant:

- Sublet work control (i.e. where the practice is subletting part of its services. Generally it is better to avoid this and arrange a direct appointment of specialist consultants e.g. for surveying).
- Organizational diagrams on jobs where the contractual relationship of organizations is complex.

It is usual to keep the master (i.e. controlled copy) of the JQP on file and to require all signing off to be done on this copy. Distributed copies can be on an uncontrolled basis or controlled. In the latter case there is a commitment to circulate all updates to holders. It should, however, be clear whether or not copies are 'controlled'.

The proforma in Appendix III can be held on a word processor to facilitate the setting up of JQPs. This method also enables them to be developed and new editions produced as the job progresses. The proforma has a section dealing with job management and guidance in development of sections for the individual professions. *Italics* are used in the proforma to indicate guidance.

To Summarize

A job quality plan must be initiated for all jobs and preparation should commence as soon as practicable after the commission is confirmed.

The JQP will contain a nucleus of all essential job specific informationi including roles and responsibilities of all main parties and the procedures to be implemented by the job team in accordance with the firm's quality system.

The master copy is kept on file and is used for the authorizing signatures and signing off of work stages. It also contains the controlled copy distribution list.

18 Auditing

Audit and management review are the parts of the quality system which are intended to ensure that the firm verifies that its quality system is implemented and continues to be implemented in accordance with the relevant parts of BS 5750. The components *Audit and Management review are* treated separately in the BS, although the clauses cross-refer to each other. They are also described separately in this section.

Audit

Definitions and Intentions of the British Standard

One of the clearest definitions of audit is given in BS 7229 Clause 2.1 (Quality audit)

> A systematic and independent examination to determine whether quality activities and related results comply with planned arrangements and whether these arrangements are implemented effectively and are suitable to achieve objectives.

The appropriate parts of the System Standard BS 5750 Part 1 are Clauses 4.17 (Internal quality audits) and a part of Clause 4.1.2.2 (Verification resources and personnel).

> 4.17: The supplier shall carry out a comprehensive system of planned and documented internal quality audits to verify whether quality activities comply with planned arrangements and to determine the effectiveness of the quality system.

Comprehensive and *planned* indicate a requirement for a programme of audits which will embrace the whole system over a given period. This is described later. The remainder of the paragraph clearly indicates the reason for audit.

> Audits shall be scheduled on the basis of status and importance of the activity.

This requires the firm to assess the relative importance to it of all the activities which come within the quality system, and to give appropriate weight to the areas which require more or less audit attention.

> The audits and follow-up actions shall be carried out in accordance with documented procedures.

A reminder that a written procedure is required to regulate the audit process.

> The results of audits shall be documented and brought to the attention of the personnel having responsibility in the area audited. The management personnel responsible for the area shall take timely corrective action on the deficiencies found by the audit (see 4.13).

This is the reporting requirement. Those directly responsible for the activity (and where the system itself is faulty, the QA manager) are to be made aware of, and correct deficiencies revealed by the audit. The reference to Clause 4.13 is to the section on Management review which takes the reporting process to the higher levels of management. This is described later in this chapter.

> 4.1.2.2: Audits of the quality system, processes . . . shall be carried out by personnel independent of those having direct responsibility for the work being performed.

The intentions of audit are therefore (in conjunction with the management review) to satisfy management that the system is operating in respect of:

- The system documentation.
- The system in operation.
- Compliance with the British Standard.

Although the term *Verification* is, perhaps confusingly, introduced into Clause 4.1.2.2, audit as a verification activity must be clearly recognized as being separate from the other verification activities required by the BS. Auditing is a neutral process of inspection by nominees of the firm's quality system (the auditors), to satisfy management that the health of the system is being maintained. Auditing should never become implicated in the quality-making process, in which 'verification' remains a matter for those who are responsible for providing the quality of the service provided to the client. There are also practical reasons for making this distinction. Only those responsible for providing the service have responsibility to the client. Auditing must never give them the opportunity to rely on some aspect of the audit process as being a part of the service. Correspondingly, auditors are neither appointed, nor necessarily professionally equipped, to seek involvement in the quality-making processes.

The Audit Process

Chapter 4 postulated the likely approach by the certification body to its own auditing process. There is not, however, any 'officially' prescribed structure or terminology which auditing should follow. The firm, if it so chooses, may develop its own process, provided it complies with Clauses 4.1.2.2 and 4.17 of BS 5750. However, the certification bodies and the auditor training organizations have between them evolved some standardization of approach and vocabulary. In addition, the publication of BS 7229 has, no doubt, consolidated movement towards a common approach. There are potential advantages to the firm in adopting this approach, particularly if certification is sought:

- Similar terminology will assist understanding and communication.
- Staff whose activities are being audited will avoid confusion caused by the different approaches of their auditors and the certification body's auditors.
- The certification body's confidence in the firm's audit procedures will be enhanced.

The description of a typical audit which follows is based on the processes and terminology adopted by most of the certification bodies.

A full audit cannot take place until:

- A procedure is in place for the audit process.
- Procedures are in place for the activity to be audited.
- The activity to be audited complies (or there is some expectation that it will comply) with the quality system.
- Appropriately trained and experienced audit resources are in place (see 'Qualities of the Auditor' later in this chapter).

However, some trial auditing is strongly recommended to familiarize both staff and auditors with the process. In such a situation, a trial audit can take place before the system is in full operation, before the full development of system documentation, and before auditors are fully trained and experienced.

An audit, as stated earlier, addresses four components:

- quality system documentation (procedures); quality system application; reporting and follow-up action.

Documentation and application

It is probable that documentation and application will be audited as one operation, because they are closely related to each other. The activity being audited is likely to be some indication of the efficacy of the procedure it is following. An audit of procedures in isolation from activities can become academic and of little help to the firm in testing its quality system.

A typical audit

An audit takes place under the overall direction of the QA manager. He must set up a comprehensive programme, and his first duty is to satisfy the *comprehensive system of planned and documented . . . audits . . .*, required by Clause 4.17 of the BS. He will write a master programme (a maximum two-year period is suggested), over which period aspects of the whole system will have been audited, having regard to the *status and importance of the activity*. See Figs 18.2 and 18.3 . . . 'master programme' at the end of this chapter. These figures indicate a 52-week programme; if a two-year programme for the whole system were adopted two such sheets would be required. It must be stressed that the activities shown represent how one firm has summarized its quality system activities to respond to its own system and compliance with the relevant parts of BS 5750 Part 1. These activities need not necessarily (and probably will not) represent the activities to be adopted by every firm. How the QA manager delegates these aspects to individual auditors will depend on the size of the firm, and in particular whether it has more than one office. For a single small office the master programme itself may well be used for this purpose. Larger or multiple offices with delegated quality management functions may write their own programmes. This will be subject to the QA manager's satisfaction that all the programmes cover separately or together the master programme. However each firm addresses this process, the auditor should prepare his own audit programme to develop his audit activities in the appropriate detail for the part of the firm he is to audit.

It is normal for the auditor to decide how he will approach the aspects he has to audit. As the man on the ground he will have a broad knowledge of the current and proposed activities of the firm or office. Most (but not all) auditing in a design organization will be carried out on the activities of job teams.

The auditor must prepare himself for the audit. The checklist (Fig. 18.4), at the end of this chapter, shows one method of approach. It identifies the level of audit (i.e. system or job) and the nature of the activity to be audited. There is space for him to indicate whether there was compliance or noncompliance, the names of those audited, and his comments. This form is also useful as an *aide-mémoire* to the auditor when he is preparing noncompliance notes (NCNs) and the audit report (these terms are defined later in this section). The auditor will also find it helpful to develop his own 'question sets', model questions or prompts derived from quality system procedures, on which he will question auditees.

Auditors should be encouraged to develop their own question sets, which comprise the model questions or 'prompts' on which they will plan their audits. They need to be thoroughly conversant with the system procedures relevant to the activities to be audited.

The auditor should give the manager of those who are to be audited (e.g. a group of design teams or the office library) notice of the parts of the system he intends to audit, and the timing he has selected. This helps to ensure that the appropriate personnel will be present.

The audit will commence with an opening meeting, when the auditor meets management to select the jobs he deems most appropriate for the parts of the system he has selected for audit. Typically in a design office with several job teams he will select from jobs which have reached certain stages (if one of the activities he is to audit is design control, he will need to select a job or jobs in the design stages).

He then visits one or more design teams. The number will depend on his programme and the scope of his audit areas. Beyond this point each audit will be different. Essentially the auditor is vigilant for aspects where the system is not being complied with; by definition if he finds nothing wrong, the system is in order and no further action is necessary.

He may cross-check between staff either to verify his findings or for consistency between staff engaged in the same, or related activities. His audit may therefore extend to, say, information sources, or archiving operations.

He will normally audit activities against the relevant quality documentation, and in design team operation particularly against the job quality plan. He may examine whether:

- Documented procedures comply with the system.
- The procedures are being complied with.
- The above comply with the intentions of quality policy.

The audit of an activity, or area of the firm's operation, may last from half an hour to half a day, depending on its extent, complexity, and importance. Typically an audit visit to a large job team, where several activities will be audited, may last up to a day. The audit time per person should not normally exceed 45 minutes. Auditing usually comprises a series of spot checks, and only rarely would a complete operation be audited.

The auditor (or auditors on a large, complex audit) needs to prepare (and concur on) the form and preparation of NCNs before formally reporting the audit findings.

The audit ends with reporting. Reporting may start with a closing-down meeting, at which the auditor meets members of management he met at the opening meeting, and representatives of those whose activities he has audited. He reports orally on his findings, and if he finds non-conformance he issues NCNs. There are two categories:

- An 'ongoing improvement' for noncompliances of a minor nature, for which evidence of correction can await his next visit, or be demonstrated by, say, a confirmatory memorandum.

- A 'hold-point'. This is issued where a serious or persistent breach of either system documentation or application is found. (See 'noncompliance note', Fig. 18.5 at the end of this chapter.) Issue of a hold-point normally requires a special return visit by the auditor to re-audit the activity for compliance and 'close-down'.

Failure in system documentation is a matter for the attention of the QA manager. Failure in application is a matter for the person undertaking the activity, or his immediate manager.

The auditor asks the senior person present at the closing-down meeting to agree the content of the NCNs and sign them.

Shortly after the audit visit, the auditor issues a general report to the management of the activities he has audited, emphasizing the more important matters (see 'audit report lead sheet', Fig. 18.6, at the end of this section). He issues a copy to the QA manager, who uses it as a basis for a management review (described later in this section). NCNs and reports of the activities audited are filed by management; copies are retained and filed by the auditor.

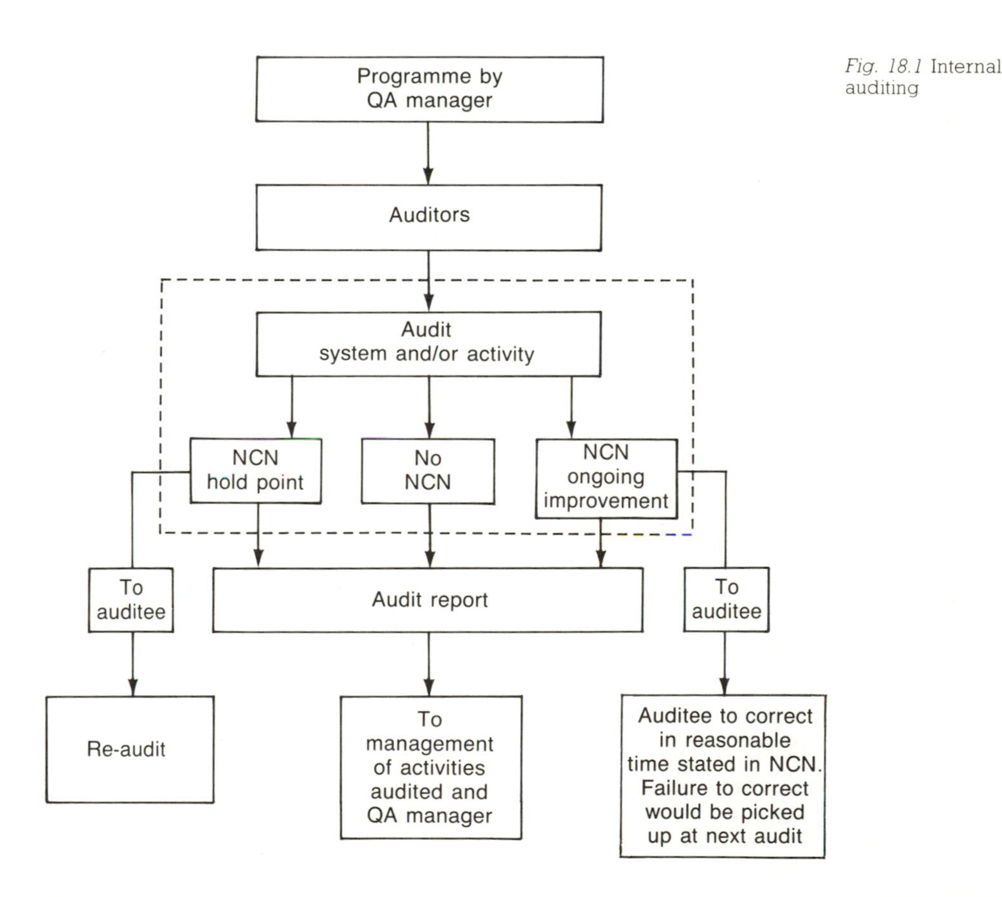

Fig. 18.1 Internal auditing

This description of a typical audit underlines the need for an auditing procedure. Auditing is a systematic and repetitive operation, and there is a need to record all aspects, from the detail of audit to maintenance of the master programme. Auditors and those audited need always to be aware of what stage has been reached in the audit process. This is particularly important when there is a need to re-audit following the issue of a hold point. Standard forms as a part of the procedure are therefore invaluable in the control, verification, and recording of the process.

The following further standard forms will be helpful to the auditor in the maintenance of proper and systematic records of the whole audit process:

- An audit report status log (see Fig. 18.7 at the end of this chapter). On this form the auditor (or in larger offices having several auditors, the QA Manager) summarizes the audit reports issued.
- A noncompliance note status log (see Fig. 18.8 at end of this chapter). The specific purpose of this form is to record and trace the history of a noncompliance.

Auditor Skills

Apart from his independence of the activity audited, an auditor can be recruited from any part of the firm's personnel, or indeed from outside the firm. For maximum benefit, however, he should be experienced, trained, and have certain qualities.

Experience, perhaps, needs no amplification. BS 5750 requires only that he be *trained, BS 7229 mentions qualification*. There is as yet no formal requirement, or even recommendation, for any given level of training. The certification bodies, which are regulated by the NACCB in the capabilities of their auditors, will require their own auditors to have attended and obtained a certificate from one of the many short courses available (typically duration 2–5 days long). They will also require their auditors to attend audits as observers before they are allowed to conduct unsupervised audits.

Firms have no obligation to train their own auditors in any particular way, but some formal training is recommended. The short courses available can give excellent basic grounding in auditing objectives, structure, awareness of BS 5750, and audit techniques. Such grounding would be tedious, if not impossible, for the firm to provide from its own resources.

There is then the question of the profession base of the potential auditor. On the one hand is the argument for an auditor whose discipline is quite different from the discipline which is undertaking the activity being audited. An auditor with such a background can bring complete detachment to the process. He can, however, waste his and the firm's time by having to learn sufficient of the activity. On the other hand there is the argument for an auditor with the same discipline. He understands the activity and indeed may have the skill to undertake it himself. He will be able to reach the truth of a situation speedily. However, because of his professional base, he can allow himself to comment on perceived shortcomings in the process, and thus be vulnerable to stepping over the responsibility line discussed earlier. (A skilled auditor may be able to assist his firm considerably by separately, i.e. not in his auditor function, and tactfully reporting shortcomings in the design itself. Clearly such a function has its dangers.)

Finally, the qualities of an auditor. Not all will be found in the same person, but the qualities to seek are:

- Ability to gain respect.
- Single-mindedness.
- Diplomacy and tact.
- Nonconfrontational.
- Ability quickly to identify the principal intentions of the procedure or activity to be audited.
- Know sufficient of the background of the activity being audited (discussed above).
- Literacy (ability to analyse, summarize speedily and with brevity).
- Persuasiveness (convincing others of the content and status of an NCN).
- Ability to determine relative importance to the firm of activities to be audited.
- Thoroughness, methodical approach to problems.
- Alertness in a repetitive process.

Management Review

In responding to Clause 4.1.3 of BS 5750 (Management review) the QA manager, normally in conjunction with the senior principal, must develop a structure and programme for periodically reviewing the whole system. This should include reporting to the senior principal, any senior management response, and the recording of the results. Clearly, feedback from certification body assessment and surveillance visits will be an important constituent of this process.

The process will probably divide into maintenance of system documentation, and audit feedback (i.e. system application). In a multi-office organization, the QA manager will also wish to include in his review strategy auditing the operation of the parts of the system he has delegated to others. Such audits must also be documented.

Fig. 18.2 Internal audit programme

MASTER PROGRAMME

PROCEDURE OR ACTIVITY TO BE AUDITED 1991 (MONTH / W/E / WK/NO)	JAN 01 02 03 04 05	FEB 06 07 08 09	MARCH 10 11 12 13	APRIL 14 15 16 17 18	MAY 19 20 21 22	JUNE 23 24 25 26	JULY 27 28 29 30 31	AUG 32 33 34 35	SEPT 36 37 38 39	OCT 40 41 42 43 44	NOV 45 46 47 48	DEC 49 50 51 52
OFFICE AUDITS												
ADMIN PROCEDURES				●							○	
MANAGEMENT RESPONSIBILITIES						●						
INFORMATION CENTRE				●							○	
TRAINING AND STAFF RECORDS				●								
ARCHIVES						●						
DOCUMENT CONTROL						●						
INTERNAL QUALITY AUDITING												
FEEDBACK CONTROL												
INSPECTION MEASURING & TEST EQUIPMENT				○→							○	
DRAWING REGISTRY				●								
JOB AUDITS (x) AUDIT NO.		(1)		(3) (4)		(6)		(7)		(9)	(10)	
JOB QUALITY PLAN(S)		●		●				●		○		
CONTRACT REVIEWS				●						○		
PROFESSION PROCEDURES				●						○		
DOCUMENT CONTROL								●				
DESIGN VERIFICATION										○		
CHECKING								●				
CHANGE CONTROL				●								
SUB-LET WORK CONTROL												
QUALITY RECORDS								●				
DRAWING REGISTRATION				●								
INSPECTION AND TESTING												
CORRECTIVE ACTION AND FEEDBACK										○		
ARCHIVING												
DATES AUDITS CARRIED OUT →		15/2		11/4 25/4		13/6		11/8				

AUDIT PROGRAMME AUTHORISED (OFFICE QA MANAGER) 31/11/90

DATE PROGRAMME REVIEWED 26/08/91

FOR ~~REVISED~~ PREVIOUS PROGRAMME SEE DOCUMENT REF NO B/IQA/001 DATE

○ PLANNED AUDIT SUBJECT

● FILLED IN WHEN COMPLETE

REF NO B/IQA/002

OFFICE BORCHESTER

INTERNAL AUDIT PROGRAMME

Fig. 18.3 Internal audit programme

MASTER PROGRAMME

PROCEDURE OR ACTIVITY TO BE AUDITED 1991 — MONTH W/E WK/NO	JAN 01 02 03 04 05	FEB 06 07 08 09	MARCH 10 11 12 13	APRIL 14 15 16 17 18	MAY 19 20 21 22	JUNE 23 24 25 26	JULY 27 28 29 30 31	AUG 32 33 34 35	SEPT 36 37 38 39	OCT 40 41 42 43 44	NOV 45 46 47 48	DEC 49 50 51 52
FIRMWIDE AUDITS												
QUALITY SYSTEM MANUAL			●									
QUALITY PROCEDURES			●						○			
MANAGEMENT RESPONSIBILITIES					●							
TRAINING AND STAFF RECORDS					●							
INTERNAL QUALITY AUDITING			●						○			
DOCUMENT CONTROL					●							
ARCHIVES									○			
FEEDBACK CONTROL												
⊛ AUDIT NO.			②		⑤				⑧			
DATE AUDITS CARRIED OUT →			13/3		16/5				24/9			

AUDIT PROGRAMME AUTHORISED (QA DIRECTOR) 31/11/90

DATE PROGRAMME REVIEWED 26/08/91

FOR PREVIOUS PROGRAMME SEE DOCUMENT REF NO B/IQA/001 DATE

REF NO B/IQA/002

FIRMWIDE INTERNAL AUDIT PROGRAMME

Fig. 18.4 Checklist

ITEM REF	CATEGORY REF* S	CATEGORY REF* J	PROCEDURAL REQUIREMENT	ACTIVITY COMPLIANCE* C	ACTIVITY COMPLIANCE* N/C	NAME(S) OF AUDITEE(S)	COMMENTS (ENTER NON COMPLIANCE REASON/SYSTEM IMPROVEMENT RECOMMENDATION)
B/915		✓	Contract Review			A Trophy	
			Stage job reached				RIBA F
			Conditions of Engagement				
			See JQP r Conditions	✓			JQP seen latest Conditions letter 13.8.90. No formal Conditions.
			Have reviews taken place – frequency? – records? – client agreement		✓		Several matters outstanding. No formal Client Agreement to date. Review at next audit – Should be Memorandum of Agreement by then.
			How have Auditees been reviewed?	✓			

*TICK EITHER

S - SYSTEMS AUDIT
J - JOB AUDIT

C - COMPLIANCE
N/C - NON COMPLIANCE

AUDIT REPORT NO B/3

PAGE 1 OF 3

CHECKLIST

Fig. 18.5 Non-compliance note

NON-COMPLIANCE NOTE

AUDITEE: A Trophy
REF NO: B 915
DATE: 11/04/91
SUBJECT OF AUDIT: Contract Review
PROCEDURE AND CLAUSE REF: QP 6/3.15–3.20
NON-COMPLIANCE NOTE NUMBER: B915/01

DETAILS OF NON-COMPLIANCE

Audit revealed several changes to service agreed informally with Client, but no record on file. No attempt to request Client to sign Memorandum of Agreement outstanding since August 1990. No formal reviews contrary to QP6 and JQP.

HOLD POINT ☑
ONGOING IMPROVEMENT ☐
SIGNATURE(S) (AUDITOR(S)) 1 C. Leadbetter 2
NAME(S) IN CAPS 1 C. LEADBETTER 2

ACCEPTED BY AUDITEE(S): A Trophy
SIGNATURE(S) 1 A Trophy 2
NAME(S) IN CAPS 1 A TROPHY 2
SIGNATURE (OFFICE QA MANAGER): M Smith
NAME IN CAPS: M SMITH DATE: 20/04/91

CORRECTIVE ACTION PROPOSED

All informal Client exchanges to be confirmed in writing and recorded on File C.3.

Client to be pressed to sign Memorandum of Agreement.

Formal reviews to take place as QP6 and frequency as stated in JQP.

PROPOSED COMPLETION DATE: 11/05/91
SIGNATURE (~~Office~~ QA MANAGER): M SMITH
NAME IN CAPS: M SMITH
DATE: 20/04/91

PROPOSAL ACCEPTED BY AUDITOR(S): ✓
SIGNATURE (AUDITOR): C Leadbetter
NAME IN CAPS: C. LEADBETTER
DATE: 11/04/91

CORRECTIVE ACTION VERIFIED

DOWNGRADE HOLD POINT TO ONGOING IMPROVEMENT ☐
SIGNATURE (AUDITOR)
NAME IN CAPS
DATE

COMPLETE ✓
SIGNATURE (AUDITOR): C Leadbetter
NAME IN CAPS: C. LEADBETTER
DATE: 15/07/91

Fig. 18.6 Typical lead sheet for audit report

AUDIT REPORT

LEAD SHEET

REPORT NO: 1.

PAGE: 1. OF 2.

SUBJECT OF AUDIT: JQP CONTRACT REVIEW PROFESSION PROCEDURES CHANGE CONTROL. DRAWING REGISTRATION

DATE OF AUDIT: 11/04/91

TYPE OF AUDIT (TICK BOX): SYSTEM ☐ JOB ☑

OFFICE LOCATION: Borchester

FOR FOLLOW UP AUDIT STATE

PREVIOUS AUDIT NO: /

PREVIOUS AUDIT DATE: /

NAMES OF AUDITEES: A TROPHY

SUMMARY OF AUDIT

Job Quality Plan

Generally in order and a true reflection of current job position. However, indications from Control Sheet that reviews last Autumn were not frequent enough to reflect substantial changes in job administration. This should be watched in future.

Contract Review

Considerable evidence of insufficient attention to the need to record and agree changes in services to be provided. Hold Point NCN issued. Job Team has been briefed satisfactorily about changes.

Profession Procedures

Generally Job Team has possession/access of latest job procedures required by JQP. Good control by office librarian of procedures to be accessed from her. However, some complaints of difficulty in obtaining Procedure No 63 — — — — — (cont'd.)

SIGNATURE (LEAD AUDITOR): C. Leadbetter

NAME IN CAPS: C. LEADBETTER

DATE: 11/04/91

SIGNATURE (~~Office~~ QA MANAGER): M Smith

NAME IN CAPS: M SMITH

DATE: 20/04/91

Fig. 18.7 Audit report status log

AUDIT REPORT NO	AUDIT TYPE* S	J	NAME OF AUDITOR	AUDIT DATE	PROCEDURE AUDITED	JOB NO (IF JOB AUDIT)	DATE AUDIT REPORT ISSUED	NO OF NCN'S ISSUED
1		✓	C LEADBETTER	11/04/91	Job Quality Plan Contract Review Profession Procedures Change Control Drawing Registration	B 915	11/04/91	1.

*TICK EITHER S - SYSTEMS AUDIT
J - JOB AUDIT

AUDIT REPORT STATUS LOG

Fig. 18.8 Noncompliance note status log

NCN SERIAL NO	JOB NO (IF JOB NCN)	NCN ISSUED TO	AUDIT REPORT NO AND DATE	TYPE OF NCN*		NAME(S) OF AUDITOR(S)	AUDITEES RESPONSE DATE TO AUDITOR(S)	PROPOSED CORRECTIVE ACTION COMPLETION DATE BY AUDITEE(S)	FOLLOW UP DATE BY AUDITOR(S)	DATE NCN CLOSED BY AUDITOR(S)
				hold point	improvement					
B 915/01	B 915	A. TROPHY	01 / 11/04/91	✓		C. LEADBETTER	11/04/91	11/05/91	11/05/91	11/05/91

* TICK EITHER

NON-COMPLIANCE NOTE

STATUS LOG

Audit feedback will comprise a substantial part of this process. The QA manager will rely heavily on the audit reports (discussed earlier) for his intelligence, both for the purpose of monitoring the audit process, and as a basis for his reporting to the senior principal. The QA manager must distil from the audit reports significant areas of noncompliance whose correction he is not able to influence. These are matters for senior principal and senior management action. Periodic analysis of NCNs and audit reports for patterns of noncompliance is an invaluable source of wider feedback material. The management review itself must also be audited.

Client and Certification Body Assessment and Audit

Client and certification body audit (second and third-party, respectively) have already been discussed in Chapter 5. Nothing further can usefully be added on the subject of client audit, because the extent and type of audit will differ widely, depending on what is agreed in the contract between client and firm.

Certification body audit will follow the pattern of the typical internal audit described above. The objectives of the certification body are similar to those of the firm: to determine that the firm has a quality system which is being maintained in accordance with BS 5750 Part 1. It is then able to give approval for the period stated in its terms. There is one important aspect where the certification body's rules for assessment differ from the firm's rules for auditing. If during the initial assessment the certification body has to issue a single 'hold-point', it may not grant approval until it has re-audited and 'closed down' the 'non-compliance note'. Equally, significant noncompliance during the period of the approval may jeopardize its renewal of the approval at the end of the period.

To Summarize

BS 5750 (Clauses 4.17 and 4.1) demand controls in *Internal quality audits* and *Management review*.

Definitions of the scope of audit (as demanded by BS 5750 and suggested in BS 7229) are discussed. Audits must be comprehensive, planned, and documented to verify that the system is appropriate and is being effectively maintained.

Independence of auditor from work being audited is stressed.

The audit process is described:

- Advisability in adopting BS terminology.

- Matters which have to be in place before audit can commence.
- Procedure for auditing.
- Procedure for the activities to be audited.
- Expected compliance by activity to be audited to quality system.
- Existence of audit resources.

Trial audit is recommended.

Relationship between system documentation and application is discussed.

A typical audit is:

- Controlled by the master programme.
- Respective approach strategies by QA manager and auditor.
- Means by which auditor prepares in detail for the audit.
- The audit itself: opening meeting, interviews, closing meeting.
- The issuing of noncompliance notes.
- The results of issuing noncompliance notes.

Proposed forms through which aspects of the audit process may be controlled are described and worked examples given.

The skills required of an auditor are discussed.

Management review is described: the interface between audit and the requirement that senior management plays its part in maintaining the auditing system.

The relationship between internal auditing and certification body audit is discussed.

19 Maintaining the Quality System

The previous four chapters have described how a firm's management creates a Quality System and sets up a system of auditing. The system does not run itself and consideration should be given to maintaining it in the future.

Senior Management

The role of senior management will continue to be to decide matters of policy, to take actions to ensure the implementation of the system and to monitor the working of the system through the QA manager.

QA Manager

The QA manager's role is to:

- Establish a programme of internal audits to cover all aspects of the system at least once a year.
- Take responsibility for the recruitment, training and functioning of auditors.
- Analyse audit reports regularly.
- Report to senior management on internal auditing, usually annually (see Chapter 18: management review).
- He may also chair a group which is responsible for reviewing and updating procedures and the firm's quality manual.
- Liaise with the external certification body if one is appointed and report to senior management.

Profession Representatives

Their role is to:

- Advise staff on the implementation of the procedures.
- Collect feedback.
- Review and update procedures.

Auditors

Their role is to:

- Carry out internal audits in accordance with the audit programme.
- Carry out follow-up audits where necessary.
- Prepare audit reports.

(See Chapter 18 for more details.)

Staff

There must continue to be an awareness of the quality system at all levels of the firm together with training and motivation. Apart from working to the procedures, staff should return feedback on them to the profession representatives. This results in a dynamic system which should continue to improve with time. One way to facilitate the development of ideas for improving procedures is to form quality circles. These are small groups of staff coming together to discuss how working methods can be improved. The DTI publication *Quality Circles* gives information on this approach.

Certification Body

When a certification body is appointed, one or more of its assessors will carry out an assessment of the firm. The assessor will issue noncompliance notes covering any points of nonconformity and an audit report. When the assessor is satisfied that an acceptable standard has been attained with the documentation and implementation of it, he will forward a report to his organization and recommend approval and registration. A Certificate of Registration will then be issued. If the assessor is not satisfied a follow-up visit will be necessary. Following approval and registration the certifier will carry out regular surveillance visits, usually every six months. (See Chapter 4 for more details of the activities of the certification body.)

20 Financial Considerations

Development Costs

The costs of developing a quality system based on BS 5750 Part 1 can be summarized under the following headings:

- Consultant.
- Reference documents (a selection from those listed in the Bibliography).
- Training key personnel i.e. the QA manager, internal auditors and those responsible for developing the quality system.
- Promoting staff awareness and working to the new procedures.
- Developing the quality system i.e. the quality system manual and supporting procedures.
- Auditing.
- Assessment by an independent certification body; this is not a requirement of the BS but is generally recommended (see Chapter 5).

Consultant

Financial help is currently available from the DTI who will pay 50% (or two-thirds of the cost if in an Assisted Area or Urban Programme Area) of the cost of between 5 and 15 man days of consultancy. This is available to firms which have fewer than 500 employees.

A booklet explaining the scheme (Introducing the enterprise initiative) can be obtained from DTI by telephoning 0800 500 200. This booklet contains the addresses of regional DTI centres to which application must be made for DTI part-funded consultancy. Before funding is provided a DTI Enterprise Counsellor will visit the firm. At this visit the counsellor will either put forward the name of an appropriate consultant or invite the firm to nominate its own consultant from the list of DTI approved consultants. If funding is then agreed, Production Engineering Research

Association (PERA) or, in the north-west, Salford University Business Services Ltd will manage the consultancy for the DTI.

The cost of employing a consultant is between £350 and £500 per day as at 1992. Typically it would be reasonable to expect the following amount of time to be required:

5 to 10 days for a small firm
10 to 15 days for a medium sized firm
15 days plus for a large firm

Reference Documents

Reference documents are listed in the Bibliography in Appendix II. Those required will depend on the area of practice covered by the design firm.

Training

There are a number of training organizations such as Bywater plc who offer training courses for key personnel. For QA managers there are three-and-a-half-day courses covering QA principles and practices and auditing. These may offer a certificate as a lead assessor on satisfactory completion of an end of course examination. For persons responsible for the development of the quality system manual and supporting procedures there are two-day courses covering QA principles and practices and procedure planning and development. Finally for auditors there are two-day courses covering the principles and practice of auditing.

As an approximate guide, two-day courses are £450 residential, £375 nonresidential and the three-and-a-half-day course £765 residential, £500 nonresidential. These are 1992 prices excluding VAT. Training organizations will run in-house courses for firms but these usually require at least 12 delegates to be economically viable. The maximum number is generally 20 delegates for two presenters.

Staff awareness is best covered by a short presentation on the general principles for all staff including secretarial staff, followed by seminars for specific interest groups. A general awareness of changes to establish good practice can be achieved in about half a day. More time will be required if existing working practices are lax. Further learning can be done on the job.

Documentation

The time taken to write the quality system manual and supporting procedures will depend on the range of activities and size of the firm and the level of documentation. The RIBA in its publication 'Quality management: guidance for an office manual' estimates that working from a sound base, the necessary documentation can be written in: 200 hours for a small firm; 500 hours for a medium sized firm and 750 hours for a large firm. For this purpose RIBA group firms as follows:

> Small firm: a firm comprising one principal. It may employ a secretary/book-keeper [who may be part time] and one or two technical assistants. Essentially few people are involved in the model. Within the profession there are many such firms.
>
> Medium firm: a firm comprising one, or more commonly two principals. It may employ ten technical and two secretarial/book-keeping staff. The firm is likely to be monodisciplinary. A maximum of 20 people are involved in the model.
>
> Large firm: a firm comprising five principals. It may employ 75 technical staff plus secretarial and other support staff. The firm may be mono- or multidisciplinary. A maximum of 100 people are involved in the model.

Maintenance Costs

It is not possible to be precise on these but the following points are relevant:

- The task of maintaining procedures once established is not onerous and as any well-run organization requires some form of procedural documentation, this should not prove to be a large additional cost.
- The system should respond to feedback, some of which will indicate improvements to procedures, resulting in efficiencies and better working methods.
- The more disciplined and systematic approach should result in greater efficiency and reduce abortive work in the long term.
- There should be a reduction of errors and greater ease in tracing any that do occur.
- Provided that the system is realistic and does not impose excessive requirements there should be little additional cost involved in working to it, apart from the specific costs identified below.

Auditing

An audit programme should be established by the QA manager with an audit of each part of the system at least every 12 months. The QA manager and internal auditors usually carry out these activities as secondary functions. This is economic and keeps them in touch with mainline activities. The amount of time per annum is estimated as follows:

Internal auditors	QA manager
8 days for a small firm.	4 days for a small firm.
12 days for a medium firm.	6 days for a medium firm.
25 days for a large firm.	12 days for a large firm.

In addition time needs to be allowed for chairing meetings to review, update and co-ordinate the procedures, and for liaison with the certification body.

Certification Body

Charges vary but are estimated by RIBA in their publication *Quality management: guidance for an office manual* at January 1990 values over a five year period:

£5,500 or £1,100 per annum for a small firm.
£7,870 or £1,570 per annum for a medium sized firm.
£11,550 or £2,310 per annum for a larger firm.

21 Legal Implications of a Quality System

This chapter is largely concerned with matters of legal liability. Hypothesis is inevitable in this area because aspects of quality management have yet to be tested in the courts. Nor does this chapter pretend to any depth of legal knowledge in a subject of considerable legal complexity. The reader should therefore accept the chapter as an introduction to the subject, and, his curiosity aroused, a basis for questions he might ask of legal authorities.

Anyone carrying on business faces risk. One of the more important risks is that someone will claim against him as a result of dissatisfaction with the services he has provided. Many firms will never face this situation; those who do will be better able to protect themselves if they have anticipated and weighed this risk, and have taken due precautions. Firms should view their quality systems as a component of total risk management.

The Effect on Risk of Quality Systems Generally

Firms who invest in formal quality systems might reasonably expect that the systematic adoption and application of sound working methods will limit the risk of being claimed against, and therefore lead to greater safety in practice. The extra dimension of a quality system provides assurance to management that the right working methods are being used in the right situations. There would seem to be little argument at this point that adoption of a formal system must enhance the safety of practice; to work systematically must be safer than working unsystematically.

Many firms will have adopted the essential features of a system without having formally recognized it as a 'Quality System'. Growth and expansion call for increased formality in working methods. The decision to write a manual confirming existing policy, and calling the whole process a 'quality system' may not alter existing risk one way or the other.

However, even the most rudimentary system should require formal commitment to stated objectives: these objectives being the working to a given range of documentation, in given circumstances. And, if management is to be reassured that the system is being implemented and maintained, there will also be an audit element. Having taken these steps, and whether or not the firm makes its quality policy practice public (through its marketing, in its approaches to clients for work, or in the contracts themselves), there must be some risk that something additional appears to be being offered. A risk that one day, through the discovery of such documents, a claimant will attempt to use the fact that the firm has a formal quality system as one of the claimant's weapons. (Whether he will succeed or not is a question which remains to be answered, and the firm must balance this possibility against the protection it believes a quality system will provide).

There will be occasions, however rigorous the system, when failure to implement a formal procedure will be used as an attempt to establish causal links with alleged error. This hypothesis again has to be balanced against the equally persuasive hypothesis that the error might still have arisen, or perhaps even greater error, had there been no procedure, or system.

These hypotheses do, however, underline the need for the firm to write procedures into its system which are no more than can reasonably be accomplished.

The Particular Effects on Risk of a Quality System which complies with BS 5750: Part 1

There is no difference in legal principle between a quality system which conforms to BS 5750 and one which does not. They are all systems based on given rules. The only difference is that on the one hand the firm has adopted rules written by a nationally recognized specifying body, and on the other the rules have been drawn from elsewhere. There may be some argument to suggest that a firm might have a better defence if it showed that it was working to a national system, than if it had gone elsewhere. But there are aspects of a BS based system which, if not addressed, could be important factors in determining the effect on a firm's potential liability.

A Guarantee, or Assurance of Product Conformance?

Generally accepted wisdom is that having a system which complies with the BS legally amounts to no more than assuring a client that a quality system which complies to a given set of rules is being applied. Legally, the mere existence of a quality system does not imply a higher level of service than that which would have been due had the system not existed. However, such assurance may be of little comfort to a firm facing litigation from an adversary who has found, or claims he has found, something in the BS through which he will claim that the firm had warranted some higher level of service in its contract of engagement. We will now explore some of the terms of the BS which might give grounds for such assumptions.

It remains to be seen whether some of the terms of BS 5750 are less precise than they might be, if attempts were made to use them to establish liability.

Clause 1.1 (scope)

> This . . . standard specifies quality system requirements for use where a contract between two parties requires the demonstration of a supplier's capability to design and supply product.

Two points arise from this extract. Firstly, the BS appears to limit its intentions as a Quality Systems Standard to its application where the contract requires *demonstration of the supplier's capability*. We shall return later to this statement in another context, but it gives some protection to a firm which can show that adoption of the Standard was not a contractual requirement. Secondly, the phrase *demonstration of capability* does not go so far as to assure the client that he will receive any given level of service, but it may establish a link between the quality system and 'product conformance', perhaps beyond the firm's intentions that *demonstration* be no more than a link between quality system and management of a process: which is much safer.

Subclause 4.2 (Quality system):

> The supplier shall establish and maintain a documented quality system as a means of ensuring that product conforms to specified requirements.

This is potentially more dangerous. It is widely held that a quality system by itself should not (and probably cannot) be a *means of ensuring that product conforms to specified requirements*. In fact, nothing can ensure the perfection of service which this quotation might imply. There is of course the more innocent alternative intention that the quality system is one of several means of achieving quality. But those words are not used, and the wording actually used may well be argued as assurance by the firm of some higher standard than it had intended.

There are other, more innocuous references in the BS, but when taken together (as no doubt they would be), it might be argued that some form of guarantee was intended. In fact the introduction itself to the BS whilst not perhaps intended as part of its contractual content is a significant statement of intention:

> For use when conformance to specified requirements is to be assured by the supplier during several stages which may include design/development . . .

The Other 'Quality' British Standards

BS 5750 and BS 4778: ISO 8402 (quality vocabulary) are the only Standards expressly intended to be read as parts of the contract between purchaser and supplier; BS 5750: Part l directly as a Systems Standard, and BS 4778 through reference to it in Clause 3 of BS 5750.

In BS 4778 we find in Clause 3.6 that quality assurance is defined as:

> . . . actions necessary to provide adequate confidence that a product or service will satisfy given requirements for quality.

Adequate of course does not mean 'absolute' as some reassurance that this Standard does not intend to imply that any form of guaranty is intended.

(The term 'Quality Assurance' is not actually used in the content of BS 5750, but it appears in the alternative ISO title (see Fig. 3.1 for titles) and in the introduction. So this definition might be established as a possible link between the system and the BS, and thus some promise of performance.)

Clause 3.11 of BS 4778 may also be significant in its definition of *quality surveillance* as *verification . . . to ensure that specified requirements for quality are being met*. However, this term is not used in BS 5750, so it may have less consequence in the liability chain.

The other British Standards contain many clauses which may associate quality management and product conformance. They are Guidance Standards not System Standards and would also seem to have less consequence. It may be noted in passing, however, that Part 4 of BS 5750 in its commentary on Clause 4.4 of BS 5750 (Design control) asks *is the fitness for purpose affected*?

Should it worry a designer that BS 5750 Part 1 might state or imply some guarantee or assurance of product conformance? Indeed it should worry a designer if he leads his client to believe that he is owed more than the skill normally to be expected of a designer. It will certainly worry the designer's professional indemnity insurer.

In the absence of any express contrary term in a contract for professional services the courts will imply a term that the designer owes a duty to take 'reasonable skill and care'. The designer should be able to have confidence that his system will not lead a court to hold that there is any higher standard. He should particularly beware the possibility that the court will determine that the standard of 'fitness for purpose', which is that established by statute as a standard applicable to the supply of goods rather than the provision of professional services, will be applied to the design services.

Thus if water leakage is caused by the detailing of a flashing in a certain position, the court will have regard, when considering the liability of the designer, to what was reasonably to be expected of the firm. If the 'reasonable skill and care' duty is established, the firm may escape. If, however, the higher level of 'fitness for purpose' is established, the firm will be held liable irrespective of the care it took in its design.

Can the firm protect itself? No one can be sure of the way the courts will interpret the law, or construe each contract. Each dispute will have different causes and will be argued differently in each action. The firm can take certain precautions by including a statement in its contract with the client to the effect that adoption of its quality system is not intended to warrant a higher duty than the normal standards of skill and care. The firm may wish to consult its legal adviser for the actual wording to be used and the protection such words might be expected to provide. The firm might wish to go further in expressly limiting its quality system to BS 5750: Part 1, thus preventing any possible reference by an adversary to any other quality Standards. It might also consider expressly excising or modifying the doubtful expressions in the BS discussed earlier. However, the danger of elaborate qualifications is that clients will begin to wonder what the firm is seeking to hide. If it is intended that BS 5750 should never be a part of the contract, any such qualifications should appear in the quality system manual and may appear also in the firm's publicity material.

British Standard 5750: Part 1 as a Constituent of the Contract between Firm and Client

The particular apprehensions of the BS as aspects of liability issues (whether it is an express part of the contract or not) have been discussed above. It is perhaps easier for an opponent to use the BS as a means of establishing liability if the BS has been formally adopted as a part of the contract, than if the firm simply lets it be known that it has a system which complies with the BS. There remains for discussion the specific duties which a firm may owe to its client, which can be implied by the BS.

It is clear from the earlier quotation from Clause 1.1 of the BS that the BS was originally intended to be expressly a part of the contract by reference. It was intended to be a basis for demonstrating to the client a stated level of quality management. Presumably, failure to maintain the level would be a breach of contract, with corresponding remedy to the client. (BS 5882: Specification for a total quality assurance programme for nuclear installations operates in this way.) The intention is reinforced in BS 5750: Part 0: Section 0.1, Clause 8.5.1 (Tailoring) where it is advised that client and firm have regard to elements which might be deleted from the Systems Standard, or added, and that these modifications be *specified in the contract*.

Current usage of the BS suggests that it plays a much less 'contractual' role than originally intended. Clients accept the firm's quality system manual as evidence of a system, particularly if is backed by certification. Nevertheless, there remains the possibility that specific compliance with the requirements of the BS will be seen as a part of the contract between firm and client. The client might be able to exercise his right under Clause 1.1 that the firm *demonstrates* its *capability to design . . . [the] product*. There is no indication or tradition to indicate what such *demonstration* might involve; it could be anything from inspecting the manual, to carrying out full audit. If the firm feels uncertain, it is recommended that consideration be given to including a statement in either the contract or the manual limiting the client's right to a given level of inspection.

Involvement by the client in the firm's quality system, typically by client audit, may of course lead the client to demand or to influence certain working methods which may or may not be how the firm would have preferred to work. Such intervention could of course mean that the client has undertaken some share of responsibility, and therefore potential liability.

The Firm and other Contracts

The firm may become involved with the quality systems of others. The contract with the client may require the firm to make certain investigations (e.g. in the appointment of a construction contractor), or the firm may wish to consider the quality systems of its sublet firms as a part of satisfying itself that they can produce the quality it requires. In principle, whether or not an organization has a quality system should pose no particular legal problems; there are many aspects which may come under consideration.

It should be borne in mind, however, that a quality system (and particularly one based on BS 5750) makes many specific demands on an organization. If such requirements are made contractually binding (as discussed above) the firm may be tempted to become implicated in the management of the firm it is investigating. While such a step may be justified in the case of its sublet firms, the greater the firm's involvement, the less the organization's responsibilities, and so, the less its potential liability (depending, of course, on the terms of the respective contracts). This is particularly important in the case of advising in the appointment of contractors and administration of the construction contract, where the general rule should be to interfere as little as possible with their organizational practices.

It may be considered good practice to ask the contractor for method statements, but care should be exercised in responding to them. The firm may comment on them but should neither approve nor disapprove, otherwise it may inadvertently assume a part of the contractor's responsibilities.

To Summarize

There is difficulty in postulating legal implications because QA is new and therefore there are no judicial precedents. The effects in risk of legal action by having, or not having, a quality system are discussed against this background. Do formalized objectives and procedures increase or reduce risk?

The particular effects in risk which might be inferred from a BS 5750 based quality system are discussed by reference to specific clauses:

- Does compliance to BS 5750: Part 1 imply guaranty, assurance of product conformance?
- Are there links between the systems standard BS 5750: Part 1 and the guidance and vocabulary standards BS 5750: Part 4 and BS 4778, which could imply guaranty, assurance of product conformance?

The dangers to the designer in widening his duties beyond 'skill and care' and how the firm might formally limit its potential liability in respect of a BS 5750: Part 1 based system are discussed.

Advice is given on the care needed to state precisely the client's powers to inspect the firm's quality system where such inspection is to be a contractual requirement.

The need for care where the firm is to become involved in the quality systems of these organizations is discussed.

22 The International Scene

A national hierarchy in the introduction of quality related activity might typically be:

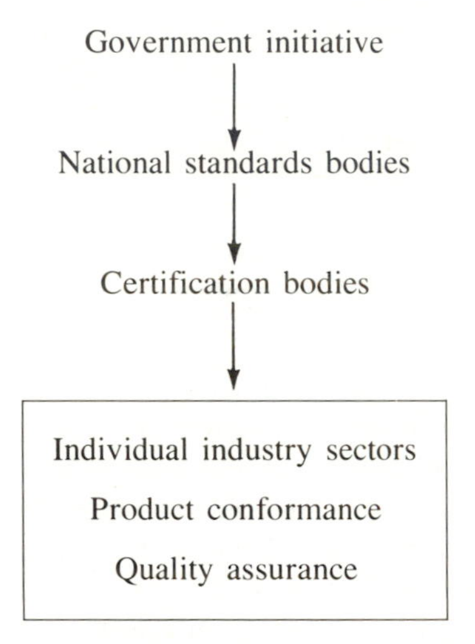

This over-simplified model postulates that governments are the prime motivating forces in seeking to improve, maintain and encourage quality in their countries. They establish a Standards base by encouraging the credibility of Standards-making bodies, and the complementary exercise of control through reputable bodies which will certify compliance with these Standards. These are the 'national controls'. Industry is then encouraged or required to comply with the Standards and undergo certification of its products. Industry as a whole is sectorized, so that each sector can identify with, support and contribute to, its own particular Standards.

Until recently, concentration focused on product conformance to a given Standard or Standards; only very recently has quality assurance started to be recognized as a separate discipline in the quality chain.

The United Kingdom has been and remains in the vanguard throughout this hierarchy; whether by far-sightedness or as some recognition that the UK as a trading nation was falling behind the rest of the world, is a matter of speculation. As can be seen from Chapter 2, the UK government led the quality drive in 1982, which promoted BSI as the prime national Standards body, and, most significantly for this book, promoted quality assurance. The manufacturing sector has continued to dominate the drive towards better quality, and the professions have had largely to tailor their own quality process formalities to industry-based Standards.

This hierarchy can be recognized in many other countries; certainly in the EC and EFTA countries, and western countries which have strong trading links with each other. Government influence in a country's quality processes varies widely; some governments exercise considerable power in the enforcement of Standards, and it is likely that their certification bodies will be an arm of government. In other countries, government exercises no more than benign encouragement, and the principal stimulus for quality lies within the private sector. It is a complicated mix, which this book cannot attempt to explore.

Such complication is nowhere more evident than in the current movement within the EC to harmonize quality related matters. However, two predictions may be made: firstly, the UK Government's national quality improvement drive, which has led to the high reputation of its Standards, and of its Standards making body, BSI, should ensure that it will exercise powerful influence in the shape of the quality process for the whole of the Community. It follows that QA will have an important place in the Community's quality requirements.

Secondly, the current negotiations between EC countries are at present preoccupied more with manoeuvring in trading advantage and disadvantage, cultural and language difficulties, and how to harmonize disparate Standards, than with promoting the practice of quality assurance as currently understood in the UK. That BS 5750 and ISO 9001 are identical Standards may owe more to the powerful international influence of BSI than to the extent of their present adoption as a base of quality systems. No-one should see them yet as total EC acceptance of QA. However, the European Council has made a resolution to promote EN 29000 (BS 5750 and ISO 9001) and EN 45012 (BS 7512: general criteria for certification bodies operating quality system certification) in an effort to encourage EC member states to promote QA and standardize its application. The Council is also encouraging member states to develop quality assessment schedules on a Europe-wide basis in the interests of removing trade barriers. (What effect such encouragement will have is difficult to predict, bearing in mind the current uncertainty in status of such schedules, as discussed in the sections on the certification bodies' schedules and quality system supplements in Chapter 4.)

A CEBI/CEDIC status report on quality assurance in European countries, Consulting Engineering Services,* largely confirms the above, although it indicates considerable interest in France to QA for consultant engineering services.

As a broad conclusion, the Western world now recognizes the value of internationally recognized Standards of product conformance; recognition of the importance of QA is less obvious.

The European countries most advanced in the theory and practice of QA are Scandinavian. Denmark (an EC member) is probably the nearest to UK practice in its adoption of QA for the design professions. The Danish Association of Practising Architects has encouraged similar initiatives. The steps in QA are part regulatory and part advisory. They bear more than a passing resemblance to BS 5750 coverage. It is interesting that their system is client led; the building owner is responsible for ensuring that consultants and contractors adopt appropriate quality systems.

Norway and Sweden (EFTA countries) follow similar principles. Norway is probably the nearest of all the Scandinavian countries to the UK in its adoption of the ISO 9000 series, and the guidance material produced by the Norwegian Building Research Institute gives detailed advice on how to create a quality system. Adoption, however, is currently confined largely to construction contractors. There is as yet relatively little movement by the design professionals.

Sweden has also adopted ISO 9000 series. Their government demands that consultants and contractors have quality systems. However, few consultants have as yet adopted QA.

The activity of QA certification bodies in Europe is low, although the Netherlands have established a Dutch equivalent of NACCB.

Of other overseas countries, the system adopted by the US Army Corps as part of its procurement package is worth mentioning. This is a combined system for ensuring quality control and quality assurance. The Corps requires the construction contractor to manage, control and document its own and its suppliers'/subcontractors' activities; processes which are close to the intentions of BS 5750. The Corps is a major contributor to the quality related processes in planning and specifying the system. Frequent inspections are made to ensure that requirements have been complied with. While design is specifically included, it would appear that independent consultants are not required to comply with the scheme. This may be because a major part of the design activities are undertaken by the contractor, reducing the perceived need for the Corps to apply quality system requirements to the scope designer.

* *Comité Européen des bureaux d'ingénierie, comité Européen des ingénieurs conseils*, published November 1990. Obtainable from Avenue Louise, 430 BTS 12-B-1050 Bruxelles.

To Summarize

The influence of governments, national standards bodies, certification bodies and individual sectors, in how the concept of quality systems is introduced.

The special part played by UK in regard to awareness and practice in quality management, and its influence in countries with which there are UK trading links.

A summary is given of the countries known to be interested in and/or practising quality management as it is known in UK. The European Council is promoting quality management under BS 5750 but few EC countries have yet adopted it to the extent it has been adopted in UK. It has been adopted outside EC countries.

23 Total Quality Management

The expression 'Total Quality Management' (TQM) is increasingly heard. While the principal purpose of this book has been to explain the operation of quality systems based on BS 5750 Part 1, some discussion of TQM might be of interest to readers.

Definition of TQM

There is as yet no widely accepted definition of TQM, unlike the quality system activities associated with BS 5750, where the relevant 'quality' British Standards provide ample definition and indeed operational guidance.

The DTI publication Total Quality Management does not give a definition as such, but states:

> 'Total Quality Management is a way of managing to improve the effectiveness, flexibility and competitiveness of the business as a whole.'

In attempting, if not a definition, then a description of some of the activities which might be included in a TQM system, we might perhaps start with the word 'total'. This word implies that there are components of a quality system which the conventionally defined quality systems do not include. Otherwise, the term 'quality system' would be sufficient to describe whatever TQM is intended to cover.

The conventional view, internationally as well as nationally, is that systems based on BS 5750 are the norm, so it may be presumed that 'total' implies something wider than the requirements of the BS. That has been taken as the starting point for discussion.

What components of a quality system are not required by BS 5750?

Historically, as described in greater detail in Chapter 2, the BS was a response to the need to improve quality management in the manufacturing of products.

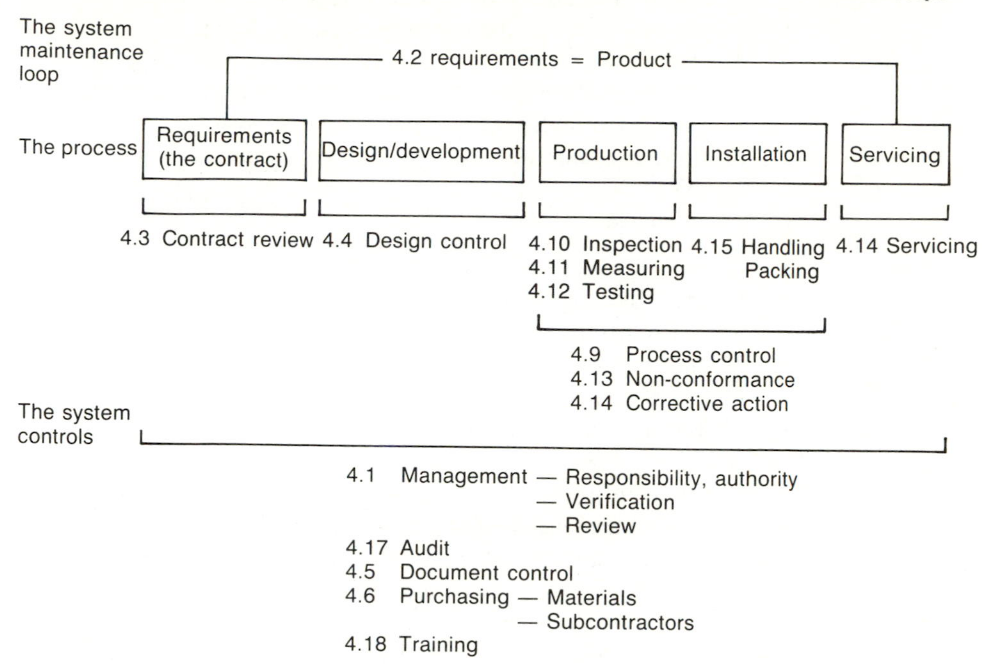

Fig. 23.1 BS 5750: Part 1: Scope

Figure 23.1 illustrates how the process is the dominant feature. Process starts with the client requirements and ends with servicing the product. The system controls address only activities directly relevant to the process. The system maintenance loop requires the system to be maintained so as to ensure that the product conforms to requirement. To over-simplify grossly for the purpose of illustrating the point, the BS was intended to be a vehicle for transferring prime responsibility for inspection of quality related aspects of the production process from purchaser to supplier. (Specifically, from government inspector to arms manufacturer.) It seems clear that it was never intended to cover activities unrelated to the manufacturing process itself.

The BS does not require the following activities to be included in a Quality System

The firm's marketing activities

- Identifying new markets.
- The detection and effects of dying markets.

- Identification of and approach to potential clients.
- Retention of goodwill of existing clients.
- Preparation and distribution of publicity material.
- The financial effects of commissions from different market sectors.
- Pursuit of specific objectives not directly related to marketing targets — e.g. design style, the seeking of design awards.
- Securing adequate resources to respond to markets pursued.

Financial controls

- Control of expenditure against income.
- Financial forecasting.
- Setting priorities for expenditure.
- Making provision for future needs.
- Means of recording and monitoring expenditure and income.
- Measuring financial performance on projects.

Purchasing

The purchases a firm makes which are not directly involved in the services produced. Although Clause 4.6 of BS 5750 is headed 'Purchasing', the precise division between direct and indirect purchases is not defined. TQM might therefore cover such matters as the firm's policy and practice in regard to maintenance of its premises or acquisition of new premises.

Subdivision of the Firm's Quality System

It is conventionally assumed for the purposes of developing a quality system that the firm is indivisible. It is assumed that it is one supplier, supplying the purchaser with services. This may be presumed from reading BS 5750 and the other Quality Standards. It might however be more realistic, particularly for larger firms, that they see themselves as several 'purchasers' and several 'suppliers', the one owing the other quality related duties. Some firms have established specialized and centralized servicing departments (e.g. office supplies, computing services). Each should perhaps establish its own quality system within the whole, recognizing internal 'clients'.

Staff capability

Clause 4.18 of BS 5750 requires training needs to be satisfied for 'personnel performing activities affecting quality'. This is normally interpreted as applying to staff directly involved in providing the service. TQM, however, might require the firm's quality system to embrace the abilities and training needs of all staff.

Staff Welfare

There is a considerable area which the conventional system does not reach, including:

- Working conditions: comfort, lighting, heating, recreation, medical facilities, safety.
- Terms of employment.
- Pension arrangements.
- Extra-mural activities (social, sports etc.).

Internal Audit

Clause 4.17 of BS 5750 specifically limits scope of audit to *verify whether quality activities comply with planned arrangements and to determine the effectiveness of the quality system*. This can be, and normally is, interpreted to mean a relatively passive, reporting operation solely concerned with the health of the system itself. TQM might encourage the more creative involvement of the auditor to include (as the DTI document suggests) 'if in the system audits it is discovered that an even better product or less waste can be achieved by changing the method or one of the materials, then a change may be effected'. This suggestion underlines a further limitation of the BS. It focuses attention on the system, rather than the intrinsic quality of the activities themselves. TQM would encourage positive attitudes by the whole workforce in their individual commitment to quality, and the goal of 'right first time'.

The Comprehensiveness of TQM

All of the above, and probably more, are components of the totality undertaken by a firm. TQM might intend them to be approached in the spirit in which the more conventional aspects of a quality system are approached. Each has to be addressed with the question 'Would the firm's overall health suffer if this activity were excluded from our quality system?' It may well be that for many firms certain aspects of their operation have been excluded for no better reason than that their inclusion is not demanded by BS 5750.

The areas described above do not include services which are traditionally provided by design firms, but are not specifically required by the BS. The most substantial activities are contract procurement and contract administration, which have been treated as normal parts of a quality system in Chapter 14. These activities are not suggested as candidates for TQM consideration, because they are directly parts of the 'product', and would, but for the manufacturing origins of BS 5750, have been as significant a part of the Standard as Clause 4: Design control.

National Recognition of TQM

Because TQM currently lacks definition by a nationally recognized authority, it may be difficult for a firm which believes it has adopted the principles of TQM either to seek a disinterested opinion on the quality of its system, to compare its system with the TQM systems of other firms, or to publicize the fact that it is operating such a system. Until system rules are published which have the authority and reputation of a British Standard, it is unlikely that any certification body would be prepared to certify its approval of TQM systems. Indeed it is doubtful whether NACCB would be empowered to accredit certification bodies who might approve such schemes.

To Summarize

TQM is as yet undefined compared with the more comprehensive definition of QA and related matters in the British Standards.

The wide range of activities which its broad ranging title implies might be included, are discussed. The assumption is made in this chapter that such matters are wider than those embraced in BS 5750: Part 1.

Possible activities are described:

- Marketing.
- Financial controls.
- Purchasing.
- Recognizing that within a firm there may be several 'purchasers' and 'suppliers', requiring more than one quality system.
- Staff welfare.
- Widening the scope of internal audit.

Appendix I
Useful Addresses

British Standards Institution (BSI),
2 Park Street,
London W1A 2BS

Tel: 071 629 9000

BSI Quality Assurance,
PO Box 375,
Milton Keynes,
Buckinghamshire MK14 6LL

Tel: 0908 220908

Bureau Veritas,
Capital House,
42 Western Street,
London SE1 3QL

Tel: 071 403 6266

Construction Industry Research and Information Association,
6 Storey's Gate,
London SW1P 3AU

Tel: 071 222 8891

Construction Quality Assurance,
Arcade Chambers,
The Arcade,
Market Place,
Newark,
Notts NG24 1UD

Tel: 0636 708700

Department of Trade and Industry,
1–19 Victoria Street,
London SW1H 0ET

Tel: 071 215 5370

Institute of Quality Assurance (IQA),
10 Grosvenor Gardens,
London SW1W 0DQ

Tel: 071 730 7154

Lloyd's Register Quality Assurance Ltd (LRQA),
Norfolk House,
Wellesley Road,
Croydon,
Surrey CR9 2DT

Tel: 081 688 6882

National Accreditation Council for Certification Bodies (NACCB),
3 Birdcage Walk,
London SW1H 9JH

Tel: 071 222 5374

RIBA Publications Ltd,
39 Moreland Street,
London EC1V 8BB

Tel: 071 251 0791

Royal Institute of British Architects,
66 Portland Place,
London W1N 4AD

Tel: 071 580 5533

Royal Institution of Chartered Surveyors,
12 Great George Street,
London SW1P 3AD

Tel: 071 222 7000

The Association of Consulting Engineers,
12 Caxton Street,
London SW1H 0QL

Tel: 071 222 6557

Yarsley Quality Assured Firms Ltd,
Trowers Way,
Redhill,
Surrey RH1 2JN

Tel: 0737 768445

Appendix II
Bibliography

Bonshor R B and Harrison H W 1982 *Quality in Traditional Housing: Volume 1 — an Investigation into Faults and their Avoidance*. Building Research Establishment Publications, Garston, Watford.

Describes a three-year study into design and construction faults in conventional two-storey housing. The report summarizes the results of the analysis of more than 90 different types of fault.

Bonshor R B and Harrison H W 1982 *Quality in Traditional Housing: Volume 2 — an Aid to Design*. Building Research Establishment Publications, Garston, Watford.

An *aide-memoire* structured by element for the design of low-rise conventionally-built housing. Linked to the 950 types of fault given in Volume 1.

Bonshor R B and Harrison H W 1982 *Quality in Traditional Housing: Volume 3 — an Aid to Site Inspection*. Building Research Establishment Publications, Garston, Watford.

Lists items for checking at ten stages of construction.

Bentley M J C *Quality Control on Building Sites*. Building Research Establishment, Garston, Watford.

A brief description of the factors which lead to defects in building and lack of quality in work.

Gregory D P *A Guide to Quality Assessment*. BSRIA, Bracknell, Berkshire.

The concept of quality assurance, and the desirability of practising a quality assurance system in business operations have now become generally well known and accepted throughout the building services industry. More recently, pressures have been applied from several sources for firms to demonstrate their commitment to quality assurance and their implementation of its principles by some form of certification or registration.

This guide presents an outline of the various ways in which such a demonstration can be achieved, either through registration in one of several similar commercial schemes or by some form of self certification. It provides guidance as to the preferred route to be taken for different situations, and explains how to go about following the selected approach.

PSA 1986 *Quality Assurance*. Property Services Agency, Croydon.

Defines what quality assurance is, how it works and why it is beneficial to all.

PSA 1987 *Quality Assurance System*. Property Services Agency, Croydon.

Quality assurance manual, a valuable source of information for companies setting up their own quality assurance schemes.

CIRIA 1985 *Sample Quality Assurance Documents*. CIRIA, London.

A selection of typical documents to indicate the general format of those which may be needed in the implementation of quality assurance schemes.

CIRIA 1988 *A Client's Guide to Quality Assurance in Construction*. CIRIA, London.

From an impartial viewpoint, briefly explains the reasons for having QA, how its introduction influences the procurement process, and what the client specifically has to do in implementing QA. Gives information on the standards relating to QA and on QA certification and accreditation.

CIRIA 1988 *Quality Assurance in Construction* (booklet accompanying video with same title). CIRIA, London.

Describes the general principles and practices of Quality Assurance and provides textual back-up for the video.

Ashford J L *Quality Management in Construction — Certification of Product Quality and Quality Management Systems*. CIRIA, London.

Examines the concept of third-party certification as it affects the construction industry and provides information on the value of certification to purchasers and suppliers. Explains both third-party certification and the role and operation of certification bodies. Concludes with an analysis of the costs and benefits of establishing a quality management system and of certification.

Oliver G B M 1990 *Quality Management in Construction — Interpretations of BS 5750 (1987) — 'Quality Systems' for the Construction Industry*. CIRIA, London.

A guide in tabular form providing interpretations of the clauses of BS 5750 for those engaged in: design and design related work; construction/installation work and monitoring of construction/installation work.

O'Brien J J 1989 *Construction Inspection Handbook; Quality Assurance and Quality Control*. 3rd edn, Van Nostrand Reinhold, New York.

Introduces the concepts of quality control and quality assurance and is organized in the same format used in building construction specifications. Incorporates all recent developments as they relate to construction.

CIOB 1987 *Quality Assurance in Building*. CIOB, Ascot, Berkshire.

Aimed at contracting organizations, this paper provides a basis for a policy of quality in building.

CIOB 1989 *Quality Assurance in the Building Process*. CIOB, Ascot, Berkshire.

Taking as its starting point the decision to build, the quality assurance implications are discussed at each stage up to and including the building's operation and maintenance. The main implications are identified for each of the parties involved throughout a typical building project, with a section being allotted to each stage.

BSI 1990 *BSI Handbook 22: Quality Assurance* 4th edn, BSI, London.

Digest of nine BS Codes on the quality theme including BS 5750.

Ashford J L 1989 *The Management of Quality in Construction*. E&FN Spon, London.

Principles and benefits of quality management by a leading practitioner. General guide plus worked model for quality system standards in design, building procurement and construction.

RIBA Indemnity Research 1989 *Quality Assurance*. RIBA, London.

Digest of information on quality systems in building design firms and what registration under BS 5750 entails. Defines the scope of international standards with relevant guidance for design practices.

Association of Consulting Engineers 1988 *Quality Assurance: Guidance note with Synopses for Quality Manual and Operational Procedures*. Association of Consulting Engineers, London.

Concise advice on the content of key system documentation; although written for engineers, it is recommended reading for any building design organization about to prepare its own documentation.

RIBA 1990 *Quality management: guidance for an office manual*.

Produced in loose leaf by the RIBA Sound Practice Committee, this provides for a demonstrable assessment of quality in architectural practice. Clear, practical advice well presented.

BPF 1983 *Manual of the BPF System for Building Design and Construction*. British Property Federation, London.

A detailed management analysis of the design and building process with precise recommendations on how clients, consultants and contractors can ensure that the client gets a building of the quality he desires, on time and within budget.

Duncan J M, Thorpe B and Sumner P 1990 *Quality Assurance in Construction*. Gower, Aldershot, Hampshire.

Provides a broad-based description of the most cost-effective way for clients and companies associated with the construction industry to apply QA within their own organization. Each sector of the industry is dealt with separately.

Ferguson I 1990 and Mitchell E 1986 *Quality on Site*. Batsford, London.

Foster A 1990 *Quality Assurance in the Construction Industry*. Hutchinson, London.

Some sectors of the industry have effectively applied QA for many years and it is now being considered seriously for widespread use in the construction industry. This book brings together a number of issues of current interest, and presents a well-rounded case for QA using both strategic and practical arguments.

Oliver G B M 1990 *Quality Management in Construction — Interpretations of BS 5750 (1987) — 'Quality Systems' for the Construction Industry*. CIRIA (SP74), London.

A guide in tabular form providing interpretations of the clauses of BS 5750 for construction professionals.

O'Reilly J 1987 *Better Briefing Means Better Buildings*. Building Research Establishment Publications, Garston, Watford.

If clients are to be satisfied with the buildings they procure, they must be involved in the overall management of the process of procurement. This BRE booklet presents a framework for briefing in the form of a checklist, approaches to project organization and overall management.

Atkinson G 1987 *A Guide through Construction Quality Standards*. Van Nostrand Reinhold (UK), Wokingham, Berkshire.

Provides the professional with a concise up-to-date guide to the complexities of quality assurance systems and regulations. It reviews: quality in construction; quality assurance schemes; building standards and regulations; product testing and certification.

Stebbing L 1986 *Quality Assurance: The Route to Efficiency and Competitiveness*. Ellis Horwood, Chichester, Sussex.

The author covers all activities through design, procurement, manufacture and installation, as well as service industry applications. He defines the contents of a quality programme and a quality plan, leading to the development of a quality manual and the supporting procedures (including auditing). Further chapters deal with computer software control and quality circles, whilst quality assurance terminology is carefully defined and many case histories given.

Haverstock Associates 1989 *Quality Management System Manual*.

It includes supporting procedures and working instructions as written for a small to medium size firm of Architects. It is a good example of a set of quality documentation.

Beaven L, Cox S, Dry D and Males R 1988 *Architect's Job Book*, 5th edn, RIBA Publications Limited, London.

(Volume 1: Job Administration; Volume 2: Contract Administration Job Record)

Plan of Work for Design Team Operation. RIBA Publications Limited, London.

Architect's Handbook of Architectural Practice. 5th edn, RIBA Publications Limited, London.

Cornick T 1990 *Quality Management for Building Design*,Butterworth Architecture, Guildford, Surrey.

Stage by stage guide to quality management.

NACCB 1991 *Directory of Accredited Certification Bodies*, NACCB, London.

DTI 1991 Department of Trade and Industry Quality Assurance Register, HMSO, Norwich.

Department of Trade and Industry Publications (obtainable from Mediascene Ltd, PO Box 90, Hengoed, Mid-Glamorgan CF8 9YE, Tel: 0443 821877)

Total Quality Management — A Practical Approach

A way of managing to improve effectiveness, flexibility and competitiveness of the business as a whole.

Leadership and Quality Management — A Guide for Chief Executives

A practical guide to help chief executives to focus on the challenge faced by managers.

The Quality Gurus — What Can They Do For Your Company?

An insight into the philosophies of eight major gurus.

The Case for Costing Quality

Case studies showing how some UK companies have tackled the problem.

The Case for Quality

How some top UK firms have used quality management in boosting competitiveness.

BS 5750/1S0 9000 : 1987 — A Positive Contribution to Better Business

How the BS applies to small and large firms and can lead to independent assessment by a third party certification body.

Best Practice Benchmarking

A case study of several companies who have used this technique in the improvement of their competitiveness.

Statistical Process Control

Case studies on statistical methods of process and product control.

Quality Circles

For the chief executive, includes case studies of successful companies that have adopted quality circles.

Problem Solving for Operators

Details of problems solving material purchasable from Bristol Polytechnic (there is also a video).

Department of Trade and Industry Videos (obtainable from Mediascene Ltd, PO Box 90, Hengood, Mid-Glamorgan CF8 9YE, Tel: 0443 821877).

Introducing Harold Slater

Amusing cartoon in understanding quality management — the hard way.

The Case of the Short-Sighted Boss

Nigel Hawthorne and Geoffrey Palmer show how Sherlock Holmes handles quality management.

An Introduction to Company-Wide Quality Improvement

Explains the roles of all employees in improving quality.

The Quality Approach

Quality for supplies officers in the Health Service.

Total Quality Management — The Right Trade

Features two fictitious companies, one of which is TQM orientated and the other not.

Total Quality Management and the Chain Reactions

How TQM can prevent the chain reactions of incompetence.

In Business to be the Best

The experience of three major companies in adopting a third party BS 5750 based system.

Quality Assurance in Construction CIRIA, London 1989.

PSA Pacesetter in Quality Assurance PSA, London 1988.

Uses the best design examples to illustrate the effectiveness of the QA system and highlights the attention to detail which can make all the difference to the end use and the satisfaction of the client. Accompanied by booklet *Quality Assurance in Building Design*.

Appendix III
Proforma for Producing a Job Quality Plan

Introduction

The following JQP proforma illustrates one approach to the preparation of Job Quality Plans. It may be adapted to suit the firm's requirements and set up on a word processor. It should include references to the firm's procedures and to file references if the firm has a standard system of referencing job files. The first section on job management (the only one included in this Appendix) will be applicable to all jobs together with other sections, for the architect, C&S engineers, etc as appropriate depending on the professions included in the commission. (These are not included in this Appendix but guidance is included at the end of it.)

It may be a requirement of the quality system that JQPs cover the relevant requirements of BS 5750: Part 1: 1987.

Guidance notes included in this model JQP (in italics) should only be deleted when actioned, unless definitely not applicable to the job, so that they will remain as an aide-mémoire *until the appropriate stage is reached and the information included in the JQP.*

The job manager should ensure that sections are included in the JQP to cover each of the professions appointed to fulfil the firm's commission, and related to the stages of work covered by the JQP.

A record of changes to the JQP should be maintained on the 'Control of changes' page. They should be referenced and adequately described and markers inserted in the right-hand margin opposite changes in the JQP. These markers will be vertical lines marked thus |. They are to be deleted in any subsequent editions of the JQP.

It is important that when JQPs are assembled, each Section and each page are similarly referenced and that pages should be consecutively numbered across all sections of the JQP.

File and procedure reference numbers should be inserted.

This introduction should be deleted when the JQP is word processed.

See Chapter 17 for further information.

*The sign *** denotes a file or procedure number to be inserted.*

Job Quality Plan

The original copy of this job quality plan is a controlled document held on file ***	Job partner's signature authorizing all Sections of this edition of the job quality plan. ....
Controlled copies of this JQP have been distributed as indicated below	Job number .... Job title .... JQP edition no. ..... Date .....

List the controlled distribution below

Job no.	Date	Edition no.	Page

Contents Page

Job no.	Date	Edition no.	Page

Record/Authorization of Changes to the Job Quality Plan

Edition no.	Page no.	Date	Change description	JP's initials

Margin marks thus | in the right-hand margin show the location of changes in the current edition of the JQP.

Record changes introduced in the 2nd and subsequent editions. Identify changes with a mark in the right-hand margin. Each revision mark will be deleted when the next edition is produced.

All editions of the JQP must be authorized by the job partner's signature on the front cover and job management summary checklists.

All amendments noted above must be authorized by the job partner's (JP's) initials in the right-hand column above.

Job no. Date Edition no. Page

Change Control Procedure

Edit procedure as necessary.

Many aspects of the firms commissions are likely to be subject to change, and will be identified at the regular job management review meetings.

The person responsible for authorizing a change on any of the undermentioned items is identified below.

Item	Person responsible for authorizing change
Job quality plan	Job partner and in the case of profession matters the job profession partner
Services and fees agreement	Job partner/senior principal or his delegated nominee
Client's brief	Job partner and the design team leader
People (team members)	Respective job profession partner
Sublet work (if applicable)	Job partner and in the case of profession matters the job profession partner
Job programmes (in-house activities)	Job manager
Job management review	Job partner
Documentation control	Job manager
Design issues	Respective job profession leaders and the design team leader
Drawings and specification (revisions)	Respective job profession leaders
	Subject to co-ordination by the design team leader

The job manager to be responsible for ensuring that key team members are informed of any changes, clearly identifying the change to the job quality plan and its associated job documentation and ensuring that the appropriate management actions are taken.

Job no.	Date	Edition no.	Page

JQP — Job Management Section

Job description
Include a brief but concise description of the job

Services and fees agreement
See file *** for full details of discussions, meetings and correspondence in respect of professional services to be given, and expenses and fee arrangements.

The services and fees agreement *has/has not been concluded and authorised by the job partner. **delete as appropriate.*

Client's brief
See file *** for full details of client briefing information.
A summary sheet of briefing information may be included for easy reference on file.

Job partner
The duties of the job partner will be as defined in the procedure ***.
If they will vary from this — specify here.

Job secretary
The duties of the job secretary will be as defined in the Office Administration Procedures.
If they will vary from this — specify here.

Job team

Enter names of team members against functions. Where functions may be relevant, but no name has yet been identified mark thus + as an aide-mémoire *for future action. Where functions will not be involved, delete.*

DTL denotes design team leader (Insert DTL against the profession leader exercising this role)

Insert names

Job partner	....
Job manager	....
Job architect partner	....
Job C&S engineer partner	....
Job building services partner	....
Job QS partner	....
Job architect	....
Job C&S engineer	....
Project engineer	....
Job QS	....
Job M&E QS	....
Job landscape architect	....
Job interior designer	....
Job planner	....
Job secretary	....
Resident site architect	....
Resident site C&S engineer	....
Resident site BS engineer	....
Resident site QS	....
Clerk of works building	....
Clerk of works building services	....

Delete as appropriate

External Organizations

List names and addresses of all external organizations, telephone numbers, fax numbers, also key people participating in the job.

Firm's name and address	Telephone no/Fax	Contact name(s)
Client	....	....
Funders	....	....

Funding agents	....	....
Building users	....	....
Project manager	....	....
External consultants *define scope of their service*	....	....
Local authorities	....	....
Statutory authorities	....	....

External Organizations

List names and addresses of all external organizations, telephone numbers, fax numbers, also key people participating in the job.

Firm's name and address	Telephone no/Fax	Contact name(s)
Other consultants or specialists *Define scope of their service*	....	....
Main contractor	....	....
Major subcontractors	....	....
Major suppliers	....	....
Others *Add any other organizations involved in the job*	....	....

Organization diagrams

Insert organization diagram(s) which illustrate the contractural and management relationships between the client, the firm and external organizations, or as an alternative, describe.

Diagram(s) when illustrated should also indicate lines of responsibility and communications, and highlight any unusual relationships.

Job no.	Date	Edition no.	Page

Sublet work
*Define those parts of the profession service to be sublet below. Set up a brief for any sublet work and control procedures in accordance with procedure ***.*

Job plan of work
*See file *** for details of the overall Job Plan of Work.*

Job programmes
*See file *** for details of the overall job programme and detailed design and production programmes. Job manager to verify that profession partners allocate adequate resources to achieve programme dates.*

*Consult procedure *** for checklist of items to be covered by programmes.*

Job partner meetings
Job manager to ensure that attendance, timing and recording of job team meetings is effectively maintained.

The following meetings will be held; basic agendas for these meetings will be as stated below.

*For model agendas for meetings, see procedure ***.*

Documentation control
*Documentation control will be maintained in accordance with the procedures set out in procedure ***. State who will deal with mail in the absence of the job manager.*

If the firm's standard procedures are not to be adopted, define alternative procedures to be used for correspondence and drawings and drawing registry and include in job quality plan.

Classification
*Classification of files and drawings to be in accordance with procedure *** — state if alternative classification system is to be used.*

Files: see file *** for current list of elemental and nonelemental files. List to be issued to job partner, job manager, design team leader, profession leaders, job secretary.

Drawing production
Define policy on the use of co-ordinated project information codes, computer aided designs and any other production techniques.

Job no.	Date	Edition no.	Page

Management review

The job partner should hold regular management review meetings, ensure that actions are recorded on the summary sheet and minutes of meetings prepared and circulated for action.

*Consult procedure *** and set up job management review programme — Alternatively this can be the responsibility of the job manager. In which case amend the above wording.*

Time sheets and expenses

Procedures in the office administration procedures to be adopted. State any variations from standard office procedures.

Job Management Summary Checklist

SETTING UP THE JOB	Signature	Date
Set up job team and hold inception meeting		
Quality objectives — Identify client's representative		
— Client's initial brief		
— Service and fees agreement		
Job computer data sheets input		
CLOSING THE JOB	Signature	Date
Completion		
Archiving		
Feedback		
Job computer data sheets completion		

Note

As each of the above items is completed the summary checklist in the controlled copy of the JQP held on file is to be signed and dated by the job partner.

Job no.	Date	Edition no.	Page

JQP — Profession Sections

The firm should develop a section or sections of the proforma for its in-house profession or professions.

The appropriate section can then be added to the JQP according to the extent of the firm's commission.

Profession sections should cover the following:

- Task list covering the service to be provided, work stage by work stage with provision for dating and signing by the profession leader as stages are complete. The task list can as a minimum be based on the conditions of engagement e.g. RIBA Architect's Appointment or ACE Conditions of Engagement.
- A commitment to adopt the firm's standard procedures or specifying deviations.
- Additional information e.g.
  The names of persons having specific responsibility, signing architect's instructions, carrying-out checking etc.
- Statement dated and signed by the job profession partner approving the JQP and confirming that any departures from deviations from standard procedures do not conflict with BS 5750: Part 1.

Index